CALCULUS

Theory and Applications in Technology and the Physical and Life Sciences

Mathematics and its Applications
Series Editor: G. M. BELL, Professor of Mathematics,
King's College London (KQC), University of London

Statistics and Operational Research
Editor: B. W. CONOLLY, Professor of Operational Research,
Queen Mary College, University of London

Mathematics and its applications are now awe-inspiring in their scope, variety and depth. Not only is there rapid growth in pure mathematics and its applications to the traditional fields of the physical sciences, engineering and statistics, but new fields of application are emerging in biology, ecology and social organisation. The user of mathematics must assimilate subtle new techniques and also learn to handle the great power of the computer efficiently and economically.

The need for clear, concise and authoritative texts is thus greater than ever and our series will endeavour to supply this need. It aims to be comprehensive and yet flexible. Works surveying recent research will introduce new areas and up-to-date mathematical methods. Undergraduate texts on established topics will stimulate student interest by including applications relevant at the present day. The series will also include selected volumes of lecture notes which will enable certain important topics to be presented earlier than would otherwise be possible.

In all these ways it is hoped to render a valuable service to those who learn, teach, develop and use mathematics.

Mathematics and its Applications
Series Editor: G. M. BELL, Professor of Mathematics, King's College London (KQC), University of London

Artmann, B. — **The Concept of Number**
Arczewski, K. & Pietrucna, J. — **Mathematical Modelling in Discrete Mechanical System**
Arczewski, K. and Pietrucha, J. — **Mathematical Modelling in Continuous Mechanical Systems**
Bainor, D.D. & Konstantinov, M. — **The Averaging Method and its Applications**
Balcerzyk, S. & Joszefiak, T. — **Commutative Rings**
Balcerzyk, S. & Joszefiak, T. — **Noetherian and Krull Rings**
Baldock, G.R. & Bridgeman, T. — **Mathematical Theory of Wave Motion**
Ball, M.A. — **Mathematics in the Social and Life Sciences: Theories, Models and Methods**
de Barra, G. — **Measure Theory and Integration**
Bell, G.M. and Lavis, D.A. — **Co-operative Phenomena in Lattice Models Vols. I & II**
Berkshire, F.H. — **Mountain and Lee Waves**
Berry, J.S., Burghes, D.N., Huntley, I.D., James, D.J.G. & Moscardini, A.O. — **Mathematical Modelling Courses**
Berry, J.S., Burghes, D.N., Huntley, I.D., James, D.J.G. & Moscardini, A.O. — **Mathematical Methodology, Models and Micros**
Berry, J.S., Burghes, D.N., Huntley, I.D., James, D.J.G. & Moscardini, A.O. — **Teaching and Applying Mathematical Modelling**
Broun, R. — **Modern Topology**
Burghes, D.N. & Borrie, M. — **Modelling with Differential Equations**
Burghes, D.N. & Downs, A.M. — **Modern Introduction to Classical Mechanics and Control**
Burghes, D.N. & Graham, A. — **Introduction to Control Theory, including Optimal Control**
Burghes, D.N., Huntley, I. & McDonald, J. — **Applying Mathematics**
Burghes, D.N. & Wood, A.D. — **Mathematical Models in the Social, Management and Life Sciences**
Butkovskiy, A.G. — **Green's Functions and Transfer Functions Handbook**
Butkovskiy, A.G. — **Structural Theory of Distributed Systems**
Cao, Z-Q., Kim, K.H. & Roush, F.W. — **Incline Algebra and Applications**
Cerny, I. — **Complex Domain Analysis**
Chorlton, F. — **Textbook of Dynamics, 2nd Edition**
Chorlton, F. — **Vector and Tensor Methods**
Cohen, D.E. — **Computability and Logic**
Crapper, G.D. — **Introduction to Water Waves**
Cross, M. & Moscardini, A.O. — **Learning the Art of Mathematical Modelling**

Series continued at back of book

CALCULUS
Theory and Applications in Technology and the Physical and Life Sciences

R. M. JOHNSON, B.Sc.(Hons.), F.I.M.A.
Senior Lecturer, Department of Mathematics and Statistics
Paisley College of Technology

ELLIS HORWOOD LIMITED
Publishers · Chichester

Halsted Press: a division of
JOHN WILEY & SONS
New York · Chichester · Brisbane · Toronto

First published in 1987 by
ELLIS HORWOOD LIMITED
Market Cross House, Cooper Street,
Chichester, West Sussex, PO19 1EB, England
The publisher's colophon is reproduced from James Gillison's drawing of the ancient Market Cross, Chichester.

Distributors:
Australia and New Zealand:
JACARANDA WILEY LIMITED
GPO Box 859, Brisbane, Queensland 4001, Australia
Canada:
JOHN WILEY & SONS CANADA LIMITED
22 Worcester Road, Rexdale, Ontario, Canada
Europe and Africa:
JOHN WILEY & SONS LIMITED
Baffins Lane, Chichester, West Sussex, England
North and South America and the rest of the world:
Halsted Press: a division of
JOHN WILEY & SONS
605 Third Avenue, New York, NY 10158, USA

British Library Cataloguing in Publication Data
Johnson, R. M. (Roy Michael)
Calculus: theory and applications in technology, and the physical and life sciences. — (Mathematics and its applications).
1. Calculus
I. Title II. Series
515 QA303

Library of Congress Card No. 87–9293

ISBN 0–7458–0325–3 (Ellis Horwood Limited — Library Edn.)
ISBN 0–7458–0288–5 (Ellis Horwood Limited — Student Edn.)
ISBN 0–470–20898–8 (Halsted Press)

Printed in Great Britain by R. J. Acford, Chichester

Table of Contents

Preface

This textbook provides a compact introductory course in calculus which includes sufficient material for a standard one-year course at university or college. It is written as a textbook for students, who are not specialist mathematicians, on courses in all areas of Engineering, Computing Science, Physics, Chemistry, Biology and Life Sciences. Additionally the book will provide a valuable calculus course for A-level students of mathematics.

The format of the book is designed to motivate students to work in self-teaching mode; each section follows the pattern of brief theory, worked examples, problems, answers. At the end of each of the main chapters there is a number of short, but fundamental, diagnostic exercises which test the degree of understanding achieved before new material is attempted.

The initial chapter (Chapter 0) deals with the revision of essential topics from algebra, geometry and trigonometry. In Chapters 1–7 the differential and integral calculus of polynomials and trigonometric functions is presented, together with applications to rates of change, optimization problems, areas and volumes. The intention is to develop a clear understanding of the fundamentals of calculus before proceeding to the more advanced functions which are considered in Chapters 8–10. Here exponential, logarithmic, and hyperbolic functions and inverse functions are introduced together with the appropriate derivatives, integrals and related techniques of integration. In Chapter 11 applications of integration are extended to include mean values, arc length, centroids and second moments. Chapter 12 introduces numerical integration and is used to emphasize that not all functions may be integrated in terms of elementary functions. The remainder of the textbook (Chapters 13–14) includes material which may not be contained in some one-year calculus courses but serves as a satisfactory base for students intending to continue their mathematical education. The topics chosen for these chapters are infinite series and differential equations.

Applications of calculus covering a wide range of disciplines are given at appropriate places in the text. Care is taken not to emphasize applications where this would detract from the objective of achieving a clear understanding of the fundamentals of calculus. Nowhere are mathematical models *derived*; it is the author's belief that the study of calculus-related model construction should be postponed until an adequate mastery of calculus has been obtained.

Much of the material in this text has been used in various degree courses at Paisley for a number of years. The development into its present form owes a lot to helpful suggestions from many colleagues. I am indebted to Professor R. R. Burnside for providing the opportunity for this project to be undertaken.

Particular thanks are expressed to Elaine Black for the considerable task of typing and correcting the manuscript.

Finally I am grateful to the Series Editor, Professor G. Bell, to the publisher's referee, and to the staff of Ellis Horwood Ltd, for their valuable assistance and encouragement.

R. M. Johnson
Paisley, 1987

0

Prerequisites from Algebra, Geometry and Trigonometry

0.1 INTRODUCTION

Before starting a first course on calculus, it is necessary to have mastered elementary algebra, geometry and trigonometry. Some of the more important topics are reviewed briefly in this chapter; for a more detailed treatment see Sweet (1984). Readers already familiar with this material should turn directly to Chapter 1.

0.2 FUNCTIONAL NOTATION

The word *function* is used to describe a relationship which assigns to each member of a set (usually a set of numbers) a particular value. The relationship is often described by a formula.

The following are examples.

(i) $A = \pi r^2$. We say that A is a function of r since for every value of r in which we are interested (usually $r > 0$) we can work out a value for A. We use the notation $A = f(r)$ or $A = A(r)$ which means 'A is a function of r'.

(ii) $v = \sqrt{20h}$. We say that v is a function of h, since for every value of h in the set $h > 0$ we can evaluate v. We write $v = f(h)$ or $v = v(h)$.

When studying the mathematics of general functions, it is customary to use the letters f, x and y although, of course *any* letters may be used. We write $y = f(x)$ which means that 'y is a function of x' or 'y depends on x'. We use the notation $f(a)$ to denote the value of the function $f(x)$ when $x = a$.

x is called the **independent variable** and the set of values of x which apply for the particular function is called the **domain** of the function.

y is called the **dependent variable** and the set of values of y (corresponding to values of x in the domain) is called the **range** of the function.

It follows that a function consists of the complete set of *ordered pairs* $(a, f(a))$ where a is a member of the domain.

For example the function $y = x^2$, with the domain $-3 \leqslant x \leqslant 3$, has the range $0 \leqslant x \leqslant 9$. That is, the function consists of the set of ordered pairs (a, a^2) where $-3 \leqslant a \leqslant 3$.

When the domain of a function is not specified, we adopt the convention that the domain is the largest set of numbers for which the function *makes sense*. This usually means that we disallow values of x which involve either

(i) division by 0 or
(ii) the square root of a negative number.

For example the function

$$y = \sqrt{\frac{x+1}{x^2}}$$

has the domain $x \geqslant -1, x \neq 0$, and the range is $y \geqslant 0$.

0.3 GRAPH SKETCHING FOR SIMPLE FUNCTIONS

The graph of a function $y = f(x)$ is a diagrammatic representation of the set of ordered pairs $(x, f(x))$ usually, but not always, drawn with x measured parallel to a horizontal axis and y measured parallel to a vertical axis. Fig. 0.1 shows the graph of $y = x^2$ and typical *points* $(2, 4), (3, 9)$ and $(-4, 16)$ on the graph.

The ability to draw quickly the graph of a given function *without* going through the process of drawing up a table of values is known as 'sketching the graph'. It forms a valuable part of the scientist's or engineer's mathematical education.

When attempting to sketch the graph of a given function the following questions should be considered.

(a) Are there any **restrictions** on the range and domain? For example, for $y = (x + 1)/x, x \neq 0, y \neq 1$.
(b) Is there any **symmetry**? For example, $y = x^2$ is symmetrical about the y axis, or $y = 1/x$ is symmetrical about the line $y = x$.
(c) Does the graph **intersect** the **coordinate axes**? For example, $y = (x - 2)/(x + 1)$ passes through $(2, 0)$ and $(0, -2)$.
(d) Are there any **stationary points**? This will be dealt with in Chapter 6 but, for simple examples, *maximum* and *minimum* points can be found by algebraic methods (see Example 0.1).
(e) What is the behaviour of the curve near its **asymptotes**? We shall define an *asymptote* as a straight line which the curve approaches as x or y (or both) becomes

very large. For example $y = 1/x^2$ has asymptotes $x = 0$ and $y = 0$, since the point (x, y) approaches the y axis as x becomes closer to zero, and (x, y) approaches the x axis as x becomes large. We write $y \to \infty$ as $x \to 0$, and $y \to 0$ as $x \to \pm\infty$.

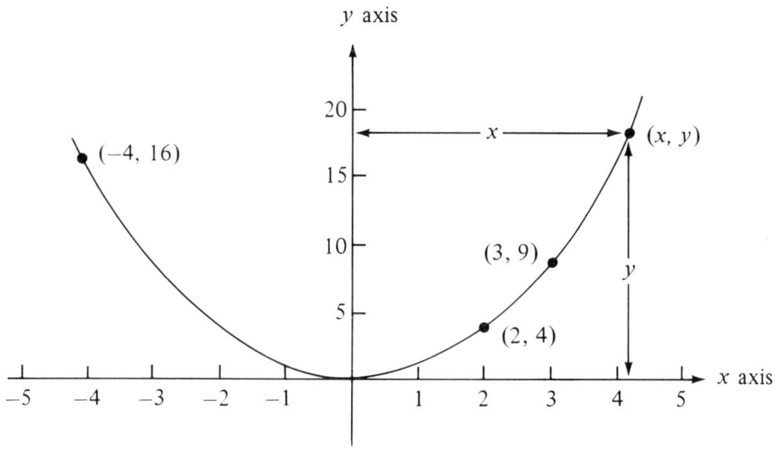

Fig. 0.1 Graph of $y = x^2$.

Example 0.1
Sketch the graph of the function

$$y = \frac{x^2 + 1}{x}.$$

(a) **Restrictions**. For the domain, $x \neq 0$. For the range, express x in terms of y, i.e. $x^2 - xy + 1 = 0$ and $x = (y \pm \sqrt{y^2 - 4})/2$. Clearly, y must satisfy $y^2 - 4 \geqslant 0$, i.e. $y \geqslant 2$ or $y \leqslant -2$.

(b) **Symmetry**. There is no obvious symmetry for this function.

(c) **Axis intersections**. The graph will intersect the x axis when $y = 0$, but $y = 0$ is not in the range of the function. The graph will intersect the y axis when $x = 0$, but $x = 0$ is excluded from the domain. Therefore there are no intersections on the co-ordinate axes.

(d) **Maximum and minimum points**. Since the range of the function is $y \geqslant 2, y \leqslant -2$, there is a *local minimum* when $y = 2$, i.e. at the point $(1, 2)$, and a *local maximum* when $y = -2$, i.e. at the point $(-1, -2)$.

(e) **Asymptotes**. For any function, two types of asymptote occur.

(i) *Vertical asymptotes corresponding to restrictions on the domain*. For this type, $x = 0$ is an asymptote and it is necessary to consider what happens as $x \to 0^-$ (from the *left* or negative side of $x = 0$) and as $x \to 0^+$ (from the *right* or positive side of $x = 0$). Consideration of $y = (x^2 + 1)/x$ shows that $y \to \infty$ as $x \to 0^+$, and $y \to -\infty$ as $x \to 0^-$ since the numerator, $x^2 + 1$, is positive for all x.

(ii) *Others found by letting* $x \to \pm\infty$. For this type of asymptote, let $x \to \pm\infty$. Rewriting the function as $y = x + 1/x$ shows that, as $x \to \pm\infty$, the graph approaches the line $y = x$, since $1/x \to 0$.

The information obtained from (a)–(e) is translated into the required sketch which is shown in Fig. 0.2.

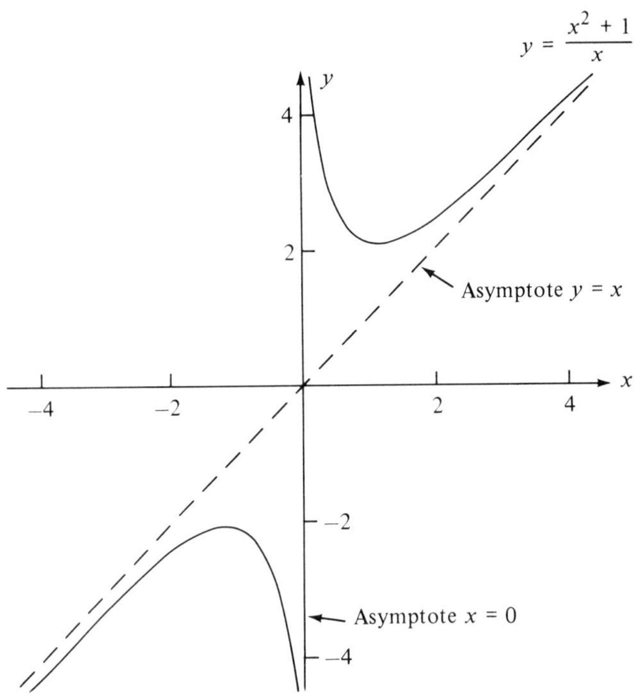

Fig. 0.2

0.4 LOGARITHMS

The function $y = a^x$, $a > 0$, is known as the **exponential function** with exponent a. In order to write the equation $y = a^x$ with x expressed in terms of y, we define the **logarithmic function** to the base a, $x = \log_a y$.

It is important to understand that $y = a^x$ and $x = \log_a y$ are simply different ways of writing the same relationship. An alternative notation for a^x is $\text{invlog}_a\, x$, the *inverse logarithm* of x.

Clearly, we have $\log_a(\text{invlog}_a\, x) = x$ and $\text{invlog}_a(\log_a y) = y$, i.e. $x = \log_a(a^x)$ and $y = a^{\log_a y}$.

Following directly from the definition of the logarithmic function and from the laws of indices ($a^x a^y = a^{x+y}$, $a^x/a^y = a^{x-y}$, $(a^x)^n = a^{nx}$) are the **laws of logarithms**:

law 1: $\log_a x + \log_a y = \log_a (xy)$,

law 2: $\log_a x - \log_a y = \log_a \left(\dfrac{x}{y}\right)$,

law 3: $n \log_a x = \log_a (x^n)$.

One value of a of particular interest in calculus is

$$e = \sum_{n=0}^{\infty} \frac{1}{n!} \approx 2.71828$$

(see Chapter 8). Common alternative notations are $\ln x = \log_e x$ and $\exp x = e^x$.

Example 0.2

Given that

$$\ln y = 2 \ln (1 + x) + 0.5 \ln (1 - x) - \ln (2 + x^2),$$

express y in terms of x.

By law 3,

$$\ln y = \ln (1 + x)^2 + \ln (\sqrt{1 - x}) - \ln (2 + x^2).$$

By law 1,

$$\ln y = \ln \{(1 + x)^2 \sqrt{1 - x}\} - \ln (2 + x^2),$$

By law 2

$$\ln y = \ln \left\{\frac{(1 + x)^2 \sqrt{1 - x}}{2 + x^2}\right\}.$$

Therefore,

$$y = \frac{(1 + x)^2 \sqrt{1 - x}}{2 + x^2}.$$

0.5 PARTIAL FRACTIONS

The method of partial fractions is used to simplify a rational function (a quotient of two polynomials) by expressing it as the sum of simpler functions. If the degree of the numerator polynomial is not less than the degree of the denominator polynomial, then the rational function must be expressed in the form;

$$\text{quotient polynomial} + \frac{\text{remainder polynomial}}{\text{denominator polynomial}}$$

the method of partial fractions is then applied to the second term. The techniques are explained in the following worked examples.

Example 0.3

Simplify the rational function

$$\frac{2x^3 + 2x^2 - 4x + 4}{x^4 - 1}.$$

Noting that the degree of the numerator polynomial is less than the degree of the denominator, the first step is to factorize the denominator into the product of linear and irreducible quadratic factors, i.e. $x^4 - 1 = (x-1)(x+1)(x^2+1)$. The rational function is then expressed as the sum of simpler functions, each containing one of the factors, as follows:

$$\frac{2x^3 + 2x^2 - 4x + 4}{x^4 - 1} = \frac{A}{x-1} + \frac{B}{x+1} + \frac{Cx + D}{x^2 + 1}. \quad (0.1)$$

Note that

(i) the numerator of each term is chosen to ensure that its degree is *one* less than the degree of the factor, and
(ii) all factors in this example are isolated (the technique for dealing with *repeated* factors is considered in Example 0.4).

The right-hand side of equation (0.1) is added to form the single term

$$\frac{A(x+1)(x^2+1) + B(x-1)(x^2+1) + (Cx+D)(x-1)(x+1)}{(x-1)(x+1)(x^2+1)}.$$

Comparing this with the left-hand side of equation (0.1) and noting that the denominators are identical, we must have

$$\begin{aligned} 2x^3 + 2x^2 - 4x + 4 = {} & A(x+1)(x^2+1) + B(x-1)(x^2+1) \\ & + (Cx+D)(x-1)(x+1). \end{aligned} \quad (0.2)$$

Equation (0.2) is in fact an *identity*, i.e. it is true for *all* values of x. By substituting appropriate values for x in equation (0.2), we obtain equations relating the constants A, B, C and D. Alternatively, other equations may be obtained by equating the coefficients of equal powers of x on both sides of equation (0.2).

The quickest procedure is usually a *combination* of these two methods. Put $x = 1$ in equation (0.2) to obtain $4 = 4A$. Put $x = -1$ in equation (0.2) to obtain $8 = -4B$. Put $x = 0$ in equation (0.2) to obtain $4 = A - B - D$. Equate coefficients of x^3 in equation (0.2) to obtain $2 = A + B + C$. Hence $A = 1, B = -2, C = 3, D = -1$ and

$$\frac{2x^3 + 2x^2 - 4x + 4}{x^4 - 1} = \frac{1}{x-1} - \frac{2}{x+1} + \frac{3x - 1}{x^2 + 1}.$$

Example 0.4
Express

$$\frac{x^5 - 3x + 1}{x(x+1)^3}$$

as the sum of a polynomial and simple rational functions.

We note that the degree of the numerator is *not* less than the degree of the denominator and it is necessary to find a *quotient* $Q(x)$ and a *remainder* $R(x)$ such that

$$\frac{x^5 + 3x + 1}{x(x+1)^3} = Q(x) + \frac{R(x)}{x(x+1)^3},$$

where the degree of $R(x)$ is less than 4. $Q(x)$ and $R(x)$ are obtained by algebraic long division or by balancing the coefficients of equal powers of x in the identity

$$x^5 - 3x + 1 = (x^4 + 3x^3 + 3x^2 + x)Q(x) + R(x).$$

Thus, $Q(x) = x - 3$, $R(x) = 6x^3 + 8x^2 + 1$ and

$$\frac{x^5 - 3x + 1}{x(x+1)^3} = x - 3 + \frac{6x^3 + 8x^2 + 1}{x(x+1)^3}. \tag{0.3}$$

The method of partial fractions is now applied to the **second term** on the right-hand side of equation (0.3) as follows:

$$\frac{6x^3 + 8x^2 + 1}{x(x+1)^3} = \frac{A}{x} + \frac{B}{x+1} + \frac{C}{(x+1)^2} + \frac{D}{(x+1)^3}. \tag{0.4}$$

Note that

(i) the factor $x + 1$ which is repeated three times gives rise to *three* terms in the simplification, and
(ii) in any simplification of partial fractions the number of constants to be determined is equal to the degree of the denominator polynomial (four in this example and four in Example 0.3).

The right-hand side of equation (0.4) is added to form the single term

$$\frac{A(x+1)^3 + Bx(x+1)^2 + Cx(x+1) + Dx}{x(x+1)^3}.$$

Note that the *same denominator* as in the original expression has been used. It is in fact the lowest common multiple of the terms x, $x + 1$, $(x + 1)^2$ and $(x + 1)^3$. Comparing the above term with the left-hand side of equation (0.4), we obtain the *identity*

$$6x^3 + 8x^2 + 1 = A(x+1)^3 + Bx(x+1)^2 + Cx(x+1) + Dx. \tag{0.5}$$

Put $x = 0$ in equation (0.5) to obtain $1 = A$. Put $x = -1$ in equation (0.5) to obtain $3 = -D$. Equate coefficients of x^3 in equation (0.5) to obtain $6 = A + B$. Equate coefficients of x^2 in equation (0.5) to obtain $8 = 3A + 2B + C$. Hence $A = 1$, $B = 5$, $C = -5, D = -3$ and

$$\frac{x^5 - 3x + 1}{x(x+1)^3} = x - 3 + \frac{1}{x} + \frac{5}{x+1} - \frac{5}{(x+1)^2} - \frac{3}{(x+1)^3}.$$

0.6 COORDINATE GEOMETRY OF THE STRAIGHT LINE AND THE CIRCLE

Many applications of calculus require knowledge of the elementary geometry of the straight line and the circle. In this section the main features of this geometry are listed for reference purposes.

(i) The **distance** between two points with coordinates (x_1, y_1) and (x_2, y_2) is given by

$$d = \sqrt{(x_2 - x_1)^2 + (y_2 - y_1)^2}.$$

(ii) The **gradient** of the line joining two points with coordinates (x^1, y^1) and (x^2, y^2) is given by

$$m = \frac{y_2 - y_1}{x_2 - x_1}.$$

Note that m is the tangent of the angle which the line makes with the *positive* x axis.

(iii) The **equation of the straight line** of gradient m passing through the point (x_0, y_0) is

$$y - y_0 = m(x - x_0).$$

(iv) The **intercept form** of the straight line equation is

$$y = mx + c,$$

which represents a line of gradient m passing through the point $(0, c)$.

(v) The equation of a straight line can be written in the **general linear form**, $1x + my + m = 0$, which represents a line *perpendicular* to the two-dimensional vector $\begin{bmatrix} l \\ m \end{bmatrix}$.

(vi) Lines parallel to the x axis are of the form $y = c$, and lines parallel to the y axis are of the form $x = k$.

(vii) Two lines with gradient m_1 and m_2 are parallel if and only if $m_1 = m_2$, and are perpendicular if and only if $m_1 m_2 = -1$.

(viii) The **equation of the circle** of radius r and with centre at the point (a, b) is

$$(x - a)^2 + (y - b)^2 = r^2.$$

(ix) The **general form** of the equation of a circle is

$$x^2 + y^2 + 2gx + 2fy + c = 0,$$

which represents a circle with centre $(-g, -f)$ and radius $\sqrt{g^2 + f^2 - c}$.

0.7 TRIGONOMETRIC FORMULAE

All the common trigonometric formulae can be derived from the pair

$$\sin(A+B) = \sin A \cos B + \cos A \sin B,$$

$$\cos(A+B) = \cos A \cos B - \sin A \sin B.$$

Those formulae most useful in techniques of calculus are listed below.

$$\cos^2 x + \sin^2 x = 1. \qquad (0.6)$$

$$1 + \tan^2 x = \sec^2 x. \qquad (0.7)$$

$$\sin 2x = 2 \sin x \cos x. \qquad (0.8)$$

$$\sin^2 x = \tfrac{1}{2}(1 - \cos 2x). \qquad (0.9)$$

$$\cos^2 x = \tfrac{1}{2}(1 + \cos 2x). \qquad (0.10)$$

$$\sin x \cos y = \tfrac{1}{2}\{\sin(x+y) + \sin(x-y\}. \qquad (0.11)$$

$$\cos x \cos y = \tfrac{1}{2}\{\cos(x+y) + \cos(x-y)\}. \qquad (0.12)$$

$$\sin x \sin y = \tfrac{1}{2}\{\cos(x-y) - \cos(x+y)\}. \qquad (0.13)$$

$$\sin x - \sin y = 2\cos\left(\frac{x+y}{2}\right)\sin\left(\frac{x-y}{2}\right). \qquad (0.14)$$

$$\cos x - \cos y = -2\sin\left(\frac{x+y}{2}\right)\sin\left(\frac{x-y}{2}\right). \qquad (0.15)$$

Problems

1. (i) Given the function $f(x) = x^3 - 3x + 1$, evaluate $f(\sqrt{3})$ and $f(-\sqrt{3})$. Does $f(x) = f(-x)$ for all x?
   (ii) Given the function $g(x) = 2x - 3$, prove that $g(a+1) - g(a) = 2$ for all values of a.

2. Find the domain and the range for the functions defined by the following equations.
   (i) $2x - y = 1$.
   (ii) $x^2 - 2y + 1 = 0$.
   (iii) $x(y-1) = 2$.
   (iv) $y = \sqrt{x^2 - 4}$.

3. Show that the function

$$\frac{x}{x-2}$$

can never take the value 1, but that this is not true for the function

$$\left(\frac{x}{x-2}\right)^2.$$

4. Find the range of the function

$$y = \frac{x^2}{x^2 - 4}.$$

5. Sketch the graphs of the following.

(i) $y = \dfrac{1}{x+1}$.

(ii) $y = \dfrac{x}{x+1}$.

(iii) $y = \sqrt{\dfrac{x}{x+1}}$.

6. The **modulus function** $|x|$ is defined as follows:

$$|x| = \begin{cases} x, & \text{when } x \geqslant 0, \\ -x, & \text{when } x < 0. \end{cases}$$

Evaluate $|0|$, $|4|$, $|-3|$, $|3-4|$. Sketch the graph of $y = |x|$. Use the graph to show that the statement $|x| < a$ is equivalent to $-a < x < a$ (a positive).

7. Sketch $y = |x-2|$ and show that, for $R > 0$, $|x-2| < R$ is equivalent to $2 - R < x < 2 + R$.

8. (i) Without using a calculator, evaluate $\log_{10} 1000$, $\log_{1000} 10$ and $\log_8(1/64)$.
(ii) Solve the following equation for y:

$$\log_{10} y = \log_{10} 8 - 2 \log_{10} 3.$$

(iii) Simplify the following expressions:

$e^{\ln x} + \ln(e^x)$,

$e^{2 \ln t}$,

$e^{t - 2 \ln t}$,

$\ln(t + 2e^t)$.

9 Express the following as the sum of partial fractions.

(i) $\dfrac{6x - 10}{x^2 - 2x + 3}$.

(ii) $\dfrac{4x}{x^3 + 1}$.

(iii) $\dfrac{s + 1}{s(s - 2)^3}$.

(iv) $\dfrac{2x^2 + 2x + 2}{(x + 1)^2 (x^2 + 1)}$.

10. Express the rational function

$$\frac{p^5 + 1}{p^2(p - 1)}$$

as the sum of a polynomial and partial fractions.

11. Find the equation in general linear form of the straight lines defined as follows.
(i) The line passing through the points $(2, -3)$ and $(1, 2)$.
(ii) The line perpendicular to $3x + 4y + 7 = 0$ and passing through $(4, 0)$.
(iii) The line passing through the point $(1, -2)$ and through the point of intersection of the lines $y = 2x + 1, y = -2x + 7$.

12. (i) Are the points $(3, 2), (6, -4)$ and $(1, 5)$ collinear?
(ii) Do the lines $y = 7x - 1, y = x + 5$ and $3x - y + 3 = 0$ have a common point?

13. Find the acute angle between the lines $y = x + 5$ and $y = 3x - 11$.

14. Find the equations in general form of the circles defined as follows.
(i) The circle passing through the points $(0, 1), (2, 2)$ and $(-1, 1)$.
(ii) The circle with diameter AB where A is $(1, 2)$ and B is $(4, 6)$.

15. (i) Find the equation of the tangent to the circle $x^2 + y^2 - 4x - 6y + 9 = 0$ at the point $(2, 1)$.
(ii) For the circle $(x - 1)^2 + (y + 2)^2 = 9$, find the length of the chord which forms part of the y axis.

16. (i) Find the equation of the circle which has its centre on the line $x - 2y + 2 = 0$ and *touches* the y axis at the point $(0, 3)$ (i.e. the y axis is a tangent to the circle).
(ii) Find the centre and radius of each of the following circles.

$$x^2 + y^2 - 8x - 6y = 0,$$

$$x^2 + y^2 - 24x - 18y = -200.$$

Deduce that the circles touch each other externally.

17. Prove the following identities:

(i) $\tan 2\theta = \dfrac{2 \tan \theta}{1 - \tan^2 \theta}$.

(ii) $\cos 4\theta = 8 \cos^4 \theta - 8 \cos^2 \theta + 1$.

18. Express the function $4 \cos 2t - 3 \sin 2t$ in the form $R \sin (2t + \phi)$ where $R > 0$ and $0 \leqslant \phi < 2\pi$. (Hint: use the $\sin (A + B)$ formula to expand $\sin (2t + \phi)$ and equate coefficients of $\cos 2t$ and $\sin 2t$.)

19. (i) Find all values of θ satisfying the equation $\sin \theta = \sin 2\theta$.
(ii) Solve the equation $\cos 2x + \sin x = 0$ for all values of x in the interval $0 \leqslant x < 2\pi$.
(iii) Find all the values of θ in $0 \leqslant \theta < 2\pi$ which satisfy the equation

$$5 \sin \left(2\theta - \frac{\pi}{4}\right) = 1.47.$$

Answers

1. (i) 1, 1, no.

2. (i) Domain, all x; range, all y.
(ii) Domain, all x; range, $y \geqslant \frac{1}{2}$.
(iii) Domain, all x except $x = 0$; range, all y except $y = 1$.
(iv) Domain, $x \geqslant 2$ and $x \leqslant -2$; range, $y \geqslant 0$.

4. $y \leqslant 0$ and $y > 1$.

6. 0, 4, 3, 1.

8. (i) $3, \frac{1}{3}, -2$.
(ii) $\frac{8}{9}$.
(iii) $2x$,
t^2,
$\dfrac{e^t}{t^2}$,
no simplification.

9. (i) $\dfrac{2}{x-3}+\dfrac{4}{x+1}$.

(ii) $-\dfrac{4}{3(x+1)}+\dfrac{4x+4}{3(x^2-x+1)}$.

(iii) $-\dfrac{1}{8s}+\dfrac{1}{8(s-2)}-\dfrac{1}{4(s-2)^2}+\dfrac{3}{2(s-2)^3}$.

(iv) $\dfrac{1}{(x+1)^2}+\dfrac{1}{(x^2+1)}$.

10. $p^2+p+1-\dfrac{1}{p}-\dfrac{1}{p^2}+\dfrac{2}{p-1}$.

11. (i) $5x+y-7=0$.
(ii) $4x-3y-16=0$.
(iii) $12x-y-14=0$.

12. (i) No.
(ii) Yes.

13. 27° approximately.

14. (i) $x^2+y^2+x-9y+8=0$.
(ii) $x^2+y^2-5x-8y+16=0$.

15. (i) $y=1$.
(ii) $4\sqrt{2}$.

16. (i) $x^2+y^2-8x-6y+9=0$.
(ii) The centre is (4, 3) and the radius is 5; the centre is (12, 9) and the radius is 5.

18. $5\sin(2t+2.214)$.

19. (i) $\theta=k\pi$, $\theta=2k\pi\pm\dfrac{\pi}{3}$, where $k=0,\pm1,\pm2,\ldots$.

(ii) $\dfrac{\pi}{2}, \dfrac{7\pi}{6}, \dfrac{11\pi}{6}$.

(iii) 0.542, 1.814, 3.684, 4.956.

1

Limits and Differentiation

1.1 INTRODUCTION

The behaviour of a function $f(x)$ as x becomes large was considered in section 0.3. For example, the function

$$f(x) = \frac{1}{1 + x^3}$$

rapidly approaches the value 0 as x becomes large (positive or negative). We say that the *limit* of $f(x)$ is zero as x tends to infinity. In this chapter, we look more closely at this behaviour and consider also the limit of a function as the independent variable tends to a finite number. Limits are important because the process of differentiation, defined in section 1.3, involves the formation of a limit.

1.2 THE DEFINITION OF A LIMIT

The first definition involves the limit of a function $f(x)$ as x tends towards a finite number a.

We say that the function $f(x)$ *tends to a limit* l *as* x *tends to* a if we can make $f(x)$ as close as we like to the number l simply by taking x sufficiently close to a (additionally $f(x)$ *remains* close to l as x becomes even closer to a).

We write $f(x) \to l$ as $x \to a$ or

$$\lim_{x \to a} \{f(x)\} = l.$$

The definition *does not imply* that $f(a) = l$. In fact, this may be true but often $f(a)$ does not exist.

For example, consider the function

$$f(x) = \frac{x^2 - 1}{x - 1} = \frac{(x + 1)(x - 1)}{x - 1}.$$

We note that $f(1)$ does not exist but, for $x \neq 1$, we have $f(x) = x + 1$. Therefore, we can make $f(x)$ as close as we like to the number 2 simply by taking x sufficiently close to 1, i.e.

$$\frac{x^2 - 1}{x - 1} \to 2 \text{ as } x \to 1$$

or

$$\lim_{x \to 1} \left(\frac{x^2 - 1}{x - 1} \right) = 2.$$

The graph of

$$y = \frac{x^2 - 1}{x - 1}$$

is rather strange; it is the straight line $y = x + 1$ with the point $(1, 2)$ missing. We shall see in section 1.3 how this rather artificial example is related to the process of differentiation.

The second definition extends the idea of a limit to the case when a is not finite.

We say that the function $f(x)$ *tends to a limit l as x tends to infinity* if we can make $f(x)$ as close as we like to the number l simply by taking x sufficiently large. ($f(x)$ must remain close to l as x becomes even larger.)

We write $f(x) \to l$ as $x \to \infty$ or

$$\lim_{x \to \infty} \{f(x)\} = l.$$

For example, consider the function

$$f(x) = 3 + \frac{1}{x}.$$

$f(x)$ can be made as close as we like to 3 by taking x sufficiently large (in particular, $f(x) - 3$ can be made less than 0.001 simply by taking x greater than 1000), i.e.

$$3 + \frac{1}{x} \to 3 \text{ as } x \to \infty$$

or

$$\lim_{x \to \infty} \left(3 + \frac{1}{x} \right) = 3.$$

Note that the graph of

$$y = 3 + \frac{1}{x}$$

has an *asymptote* $y = 3$.

A similar definition covers the case when x tends to negative infinity.

The 'wordy' definitions for $\lim_{x\to a}\{f(x)\}$ and $\lim_{x\to\infty}\{f(x)\}$ given in the preceding paragraph are considered sufficient to provide an understanding of the concept. Further study of limits would require the formal mathematical definitions which can be found in most calculus textbooks; in particular see Dunning-Davies (1982).

The following theorem is useful when calculating limits; its proof follows from the definition of a limit.

Theorem 1.1

Given that $\lim_{x\to a}\{f(x)\} = l$ and $\lim_{x\to a}\{g(x)\} = m$, then

(i) $\lim_{x\to a}\{pf(x) + qg(x)\} = pl + qm$, p and q constants,

(ii) $\lim_{x\to a}\{f(x)g(x)\} = lm$ and

(iii) $\lim_{x\to a}\left\{\frac{f(x)}{g(x)}\right\} = \frac{l}{m}$, provided that $m \neq 0$

The result

$$\lim_{x\to 0}\left(\frac{\sin x}{x}\right) = 1$$

is of particular importance and will be applied in section 1.5 to the differentiation of trigonometric functions. To demonstrate this result, try using a calculator to evaluate the function

$$\frac{\sin x}{x}$$

for successively smaller values of x (**radians!**, usually abbreviated to rad); see also Example 1.2. The statement

$$\lim_{x\to 0}\left(\frac{\sin x}{x}\right) = 1$$

is another way of saying that

$$\frac{\sin x}{x} = 1$$

approximately *when x is small*, i.e.

$$\sin x \approx x \quad \text{when } x \text{ is small.} \tag{1.1}$$

Equation (1.1) is known as the **small-angle approximation** and is useful in many practical situations where the independent variable is known to lie in an interval *near* $x = 0$. Sometimes equations which cannot be solved analytically are reduced to simple equations by application of equation (1.1). For example, given that the equation $10x^2 = \cos x$ has solutions near $x = 0$ (easily verified by sketching $y = 10x^2$ and $y = \cos x$), approximate solutions can be found as follows.

It is easily shown that the equation $100x^4 = \cos^2 x$ has the same solution set and, using $\cos^2 x + \sin^2 x = 1$ (see section 0.7), we have

$$100x^4 = 1 - \sin^2 x.$$

Applying the small-angle approximation gives

$$100x^4 + x^2 - 1 = 0,$$

i.e.

$$x^2 = \frac{-1 \pm \sqrt{401}}{200} = 0.0951$$

(rejecting the negative sign) and $x = \pm 0.3084$ are the approximate solutions. (Compare with $x = \pm 0.3087$ which satisfy $10x^2 = \cos x$ accurate to four decimal places.)

Example 1.1

Evaluate the following limits.

(i) $\lim_{x \to 2} (x^2 + 2)$.

(ii) $\lim_{x \to 2} \left(\frac{x^2 - 4}{x - 2}\right)$.

(iii) $\lim_{x \to \infty} \left(\frac{x^2}{x^2 + 2x + 3}\right)$.

(iv) $\lim_{x \to 0} \left(\frac{x + 1}{x}\right)$.

(i) The function $x^2 + 2$ causes no problems at $x = 2$; it exists and has the value 6. Clearly, $x^2 + 2$ can be made as close as we like to 6 by taking x sufficiently close to 2. Therefore,

$$\lim_{x \to 2} (x^2 + 2) = 6.$$

(ii) There is a problem for the function

$$\frac{x^2-4}{x-2}$$

at $x = 2$; the function is not defined. However, remember that $x \to 2$ does not mean that x ever reaches the value 2; x can be arbitrarily close to 2 without being equal to 2. For $x \neq 2$,

$$\frac{x^2-4}{x-2} = x+2.$$

Therefore,

$$\lim_{x\to 2}\left(\frac{x^2-4}{x-2}\right) = \lim_{x\to 2}(x+2),$$

since $x \neq 2$, i.e. the limit is 4.

(iii) One way to do this type of example is to put $x = 1/h$ and to take the limit as $h \to 0$ (i.e. $x \to \infty$). Therefore,

$$\lim_{x\to\infty}\left(\frac{x^2}{x^2+2x+3}\right) = \lim_{h\to 0}\left(\frac{1/h^2}{1/h^2+2/h+3}\right)$$

$$= \lim_{h\to 0}\left(\frac{1}{1+2h+3h^2}\right),$$

since $h \neq 0$. Thus,

$$\lim_{h\to 0}\left(\frac{1}{1+2h+3h^2}\right) = 1.$$

Note that with practice the above device becomes unnecessary. A glance at the function is enough to see that the x^2 terms dominate when x is large and the function is effectively x^2/x^2 as $x \to \infty$.

(iv) $$\lim_{x\to 0}\left(\frac{x+1}{x}\right)$$

does not exist since

$$\frac{x+1}{x} = 1 + \frac{1}{x}$$

which increases beyond the bound as x tends to zero. However, we can write

$$\frac{x+1}{x} \to \infty \text{ as } x \to 0^+$$

and

$$\frac{x+1}{x} \to -\infty \text{ as } x \to 0^-.$$

In fact, $x = 0$ is an asymptote of the graph of

$$y = \frac{x+1}{x}.$$

Example 1.2

(i) By ordering the areas of triangle OPA, sector OPA and triangle OPT in Fig. 1.1, show that

$$\lim_{x\to 0}\left(\frac{\sin x}{x}\right) = 1.$$

(ii) Evaluate

$$\lim_{x\to 0}\left(\frac{\sin 3x}{\sin 4x}\right).$$

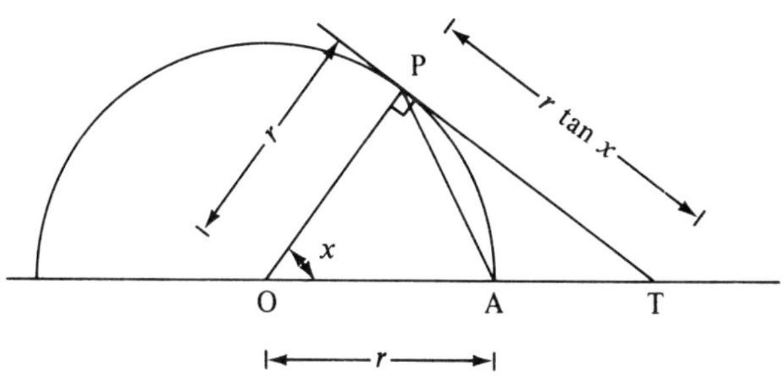

Fig. 1.1

(i) In Fig. 1.1 O is the centre of the semicircle of radius r and PT is the tangent at P. Angle POA is x rad. We have that area of triangle OPA < area of sector OPA < area of triangle OPT, i.e.

$$\tfrac{1}{2}r^2 \sin x < \tfrac{1}{2}r^2 x < \tfrac{1}{2}r^2 \tan x$$

$$\sin x < x < \frac{\sin x}{\cos x},$$

$$1 < \frac{x}{\sin x} < \frac{1}{\cos x},$$

$$1 > \frac{\sin x}{x} > \cos x.$$

(This holds for $x > 0$ *and* $x < 0$.) Now let $x \to 0$. Therefore,

$$1 > \lim_{x \to 0} \left(\frac{\sin x}{x} \right) > \lim_{x \to 0} (\cos x) = 1,$$

and thus

$$\lim_{x \to 0} \left(\frac{\sin x}{x} \right) = 1.$$

(ii) Since $x \to 0$, we have both $3x$ and $4x$ *small*. Therefore, equation (1.1) can be applied to give $\sin 3x \approx 3x$ and $\sin 4x \approx 4x$. Therefore,

$$\lim_{x \to 0} \left(\frac{\sin 3x}{\sin 4x} \right) = \lim_{x \to 0} \left(\frac{3x}{4x} \right) = \frac{3}{4}.$$

Problems

1. Evaluate $\lim_{x \to a} \{ f(x) \}$ for the following $f(x)$ and a.

(i) $f(x) = \dfrac{2x + 4}{x^2 - 2}$, $a = 3$.

(ii) $f(x) = \dfrac{2x - 4}{x^2 + 2}$, $a = 2$.

(iii) $f(x) = \dfrac{x^2 - 4}{x^2 - 5x + 6}$, $a = 2$.

(iv) $f(x) = \dfrac{(x - 2)^2}{x^2 - 4}$, $a = 2$.

2. Evaluate $\lim_{x \to \infty} \{ f(x) \}$ for the following $f(x)$.

(i) $f(x) = \dfrac{7x^7 - 7}{8x^7 + 8}$.

(ii) $f(x) = \dfrac{x^3 + 5}{2x^4 - 1}$.

(iii) $f(x) = \dfrac{x^3}{(x - 1)^2}$.

3. Given that

$$g(t) = \frac{t^2 - 9}{t^2 - 4t + 3},$$

evaluate $\lim_{t \to a} \{g(t)\}$ for the following.

(i) $a = 2$.
(ii) $a = 3$.
(iii) $a = -3$.
(iv) $a = 1$.

4. Evaluate the following.

(i) $\lim_{\theta \to 0} \left(\frac{\sin 2\theta}{\theta} \right)$.

(ii) $\lim_{t \to 0} \left(\frac{t}{\tan t} \right)$.

(iii) $\lim_{x \to \pi/2} \left(\frac{\cos x}{(\pi/2) - x} \right)$.

5. Sketch the graph of the function $y = f(x)$ where

$$f(x) = \begin{cases} 1 + x, & x < 0, \\ 2 + x, & x \geqslant 0. \end{cases}$$

Evaluate $\lim_{x \to 0^-} \{f(x)\}$ and $\lim_{x \to 0^+} \{f(x)\}$. (Note that these limits are not equal; in cases where $\lim_{x \to a^-} \{f(x)\} \neq \lim_{x \to a^+} \{f(x)\}$ we say that $\lim_{x \to a} \{f(x)\}$ does not exist.)

6. The function $f(x)$ is said to be **continuous** at $x = a$ if $\lim_{x \to a} \{f(x)\}$ exists *and* is equal to $f(a)$. (This definition implies that the graph of $y = f(x)$ can be drawn through the point where $x = a$ without removing pen from paper.)

(i) Confirm that the function defined in problem 5 is continuous for all x except $x = 0$.
(ii) Sketch roughly the graph of $y = 1/x^2$ and explain why it is not continuous when $x = 0$.

Answers

1. (i) $\frac{10}{7}$.
(ii) 0.
(iii) -4.
(iv) 0.

2. (i) $\frac{7}{8}$.
   (ii) 0.
   (iii) Limit does not exist.
3. (i) 5.
   (ii) 3.
   (iii) 0.
   (iv) Limit does not exist.
4. (i) 2.
   (ii) 1.
   (iii) 1.
5. 1, 2.

1.3 THE DERIVATIVE OF A FUNCTION

Given the function $f(x)$ we may define a related function, denoted by $f'(x)$, as follows:

$$f'(x) = \lim_{h \to 0}\left\{\frac{f(x+h)-f(x)}{h}\right\} \tag{1.2}$$

provided that the limit exists.

$f'(x)$ is called the **first derived function** of $f(x)$, or simply the **derivative** of $f(x)$. The process of forming $f'(x)$ from $f(x)$ is called **differentiation with respect to** x.

The function $f'(x)$ is in fact a measure of the *rate of change* in $f(x)$ with respect to change in x. To see this, consider Fig. 1.2 which shows the graph of $y = f(x)$ passing through the points P and Q. The coordinates of P are (x, y) and the symbols Δx and Δy represent the changes in x coordinate and y coordinate respectively between the points P and Q. Hence the coordinates of Q are $(x + \Delta x,\ y + \Delta y)$ and, clearly $\Delta y = f(x + \Delta x) - f(x)$. We define the **average rate of change** in y with respect to x in the interval between P and Q as

$$\frac{\text{change in } y}{\text{change in } x} = \frac{\Delta y}{\Delta x}.$$

This is the *gradient* of the straight-line segment PQ as defined in section 0.6. The line PQ is called a *chord* of the curve $y = f(x)$. The **instantaneous rate of change** at P in y with respect to x is obtained by creating a limiting process where Q is allowed to move along the graph towards P. An approximation to the *instantaneous* rate of change is obtained by calculating the *average* rate of change between P and Q with Q *close* to P. The approximation is improved by taking Q even closer to P and hence

instantaneous rate of change of y at P

$$= \lim_{Q \to P} (\text{average rate of change in } y \text{ between P and Q})$$

$$= \lim_{\Delta x \to 0}\left(\frac{\Delta y}{\Delta x}\right)$$

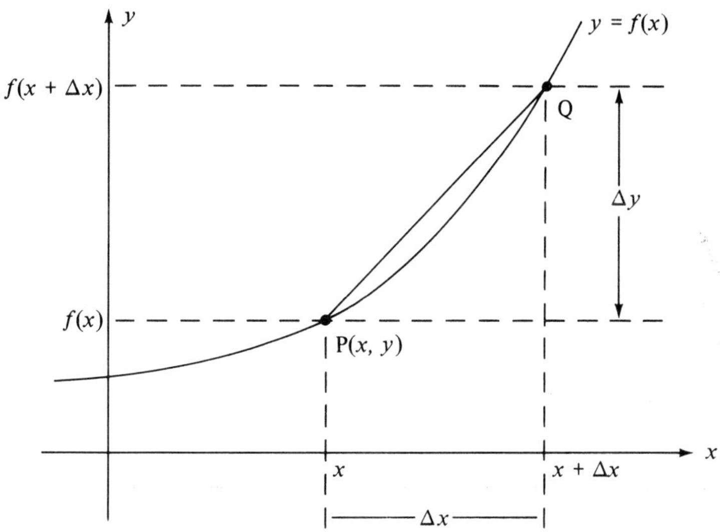

Fig. 1.2

$$= \lim_{\Delta x \to 0} \left\{ \frac{f(x + \Delta x) - f(x)}{\Delta x} \right\}$$

$$= f'(x),$$

(the symbol Δx replacing the variable h in equation (1.2)). Note that as $\Delta x \to 0$ the chord PQ approaches the *tangent line* PT, the line which *just touches* the curve at P. It follows that the gradient of the tangent line at P on the curve $y = f(x)$ is given by the *value* of $f'(x)$ at P.

An alternative notation for the derivative of the function $y = f(x)$ is $\mathrm{d}y/\mathrm{d}x$; it must be understood at this stage that $\mathrm{d}y/\mathrm{d}x$ is merely a *symbol* for the derivative and does not imply a division of $\mathrm{d}y$ by $\mathrm{d}x$.

The most common application of the concept of rate of change is **velocity** which is the *rate of change in distance with respect to time* (typical units for velocity are miles per hour (miles/h) or metres per second (m/s), i.e. distance per time). Consider the motion of a vehicle travelling in a straight line where its distance s m from a fixed point in the line is given as a function of time t s. If the vehicle is travelling at a constant velocity, then the graph of s against t is a straight line. Usually, velocity varies with time and the graph of the function $s = f(t)$ has a curved shape. Fig. 1.3 shows a typical s versus t graph for an *accelerating* situation; P and Q are points on the curve corresponding to the times t_1 and t_2. The *average* velocity in the time interval $t_1 \leqslant t \leqslant t_2$ is calculated simply as

$$\frac{\text{change in distance}}{\text{change in time}} = \frac{s_2 - s_1}{t_2 - t_1},$$

i.e. the gradient of line PQ,

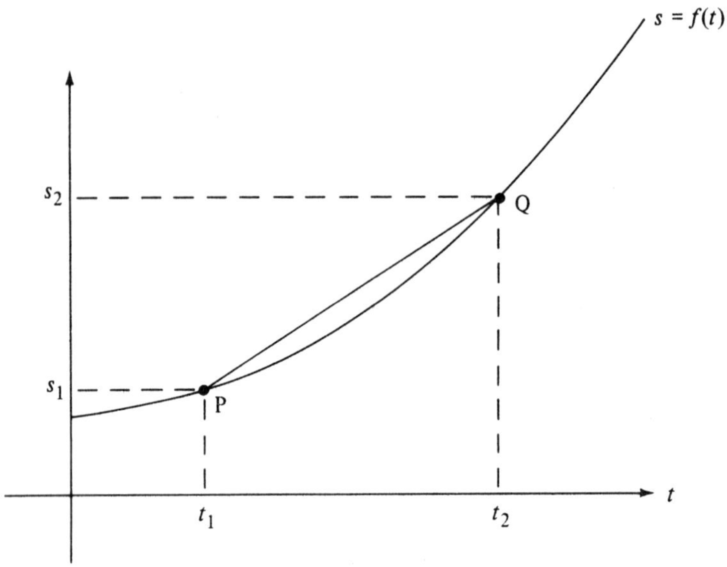

Fig. 1.3 A velocity versus time curve.

$$\frac{f(t_1+h)-f(t_1)}{h},$$

where $h = t_2 - t_1$. The *instantaneous* velocity at the instant $t = t_1$ can be obtained *approximately* by calculating the average velocity over a *small* interval starting at $t = t_1$. The *exact* instantaneous velocity is obtained by allowing the size of this small interval to tend to zero, i.e. by calculating the *limit*

$$\lim_{h\to 0}\left\{\frac{f(t_1+h)-f(t_1)}{h}\right\}.$$

Reference to equation (1.2) shows that the instantaneous velocity at time t_1 is $f'(t_1)$, the value of $f'(t)$ when $t = t_1$. It follows that the velocity v of the vehicle at *any* time is given by the value of $f'(t)$ at that time, i.e. $v = f'(t)$ or using the alternative notation

$$v = \frac{\mathrm{d}s}{\mathrm{d}t}. \tag{1.3}$$

It should be noted that the **acceleration** a, of the vehicle is defined as the *rate of change in velocity with respect to time* (typical units are metres per second/second (m/s^2)). Therefore, we have

$$a = \frac{\mathrm{d}v}{\mathrm{d}t}. \tag{1.4}$$

A further application of differentiation follows from the definition of the **gradient of a curve**. The *gradient* of the curve $y = f(x)$ at the point P(x, y) is defined as the gradient of the *tangent line* to the curve at P and is calculated by evaluating $f'(x)$ at P. It is then a simple matter to obtain the equation of the tangent line to a curve $y = f(x)$ at a given point provided that the derivative $f'(x)$ is available.

Since the gradient of the tangent line to a curve $y = f(x)$ at a point is a measure of the value of $f'(x)$ at the point, it follows that $f'(x)$ (or $\mathrm{d}y/\mathrm{d}x$), can be evaluated approximately at any point simply by drawing the tangent line to the curve and measuring its gradient. The same graphical method may be applied to obtain velocity $\mathrm{d}s/\mathrm{d}t$ from the distance versus time curve and allows an approximate velocity versus time curve to be constructed. Similarly the acceleration $\mathrm{d}v/\mathrm{d}t$ may be estimated from the velocity versus time curve.

Example 1.3

(i) Given that $f(x) = x^2$ use the definition of $f'(x)$ to prove that $f'(x) = 2x$.

(ii) Confirm that

$$z^n - x^n = (z - x) \sum_{k=1}^{n} z^{n-k}x^{k-1},$$

where n is a positive integer, and evaluate

$$\lim_{z\to x} \left(\frac{z^n - x^n}{z - x} \right).$$

Hence show that, for $f(x) = x^n$, n a positive integer,

$$f'(x) = nx^{n-1}.$$

(i) $f(x) = x^2$ and $f(x + h) = (x + h)^2 = x^2 + 2xh + h^2$. Therefore,

$$\frac{f(x + h) - f(x)}{h} = \frac{2xh + h^2}{h} = 2x + h.$$

Therefore,

$$f'(x) = \lim_{h\to 0} (2x + h) = 2x.$$

(ii) $$(z - x) \sum_{k=1}^{n} z^{n-k}x^{k-1} = (z - x)(z^{n-1} + z^{n-2}x + z^{n-3}x^2 + \ldots + x^{n-1})$$

$$= z^n - x^n,$$

which is confirmed simply by multiplication.

$$\frac{z^n - x^n}{z - x} = \frac{(z - x) \sum_{k=1}^{n} z^{n-k}x^{k-1}}{z - x}$$

$$= \sum_{k=1}^{n} z^{n-k}x^{k-1}.$$

Therefore,

$$\lim_{z\to x}\left(\frac{z^n - x^n}{z - x}\right) = \lim_{z\to x}\left(\sum_{k=1}^{n} z^{n-k}x^{k-1}\right)$$

$$= \sum_{k=1}^{n} x^{n-1}$$

$$= nx^{n-1}.$$

For $f(x) = x^n$,

$$f'(x) = \lim_{h\to 0}\left\{\frac{(x+h)^n - x^n}{h}\right\}.$$

Let $x + h = z$ and

$$f'(x) = \lim_{z\to x}\left(\frac{z^n - x^n}{z - x}\right) = nx^{n-1}.$$

Example 1.4

Use the definition of a first derivative to obtain $\mathrm{d}y/\mathrm{d}x$ when $y = \sqrt{x}$.

Defining Δy as the change in y as x changes by amount Δx, we have

$$\Delta y = \sqrt{x + \Delta x} - \sqrt{x}.$$

$$\frac{\mathrm{d}y}{\mathrm{d}x} = \lim_{\Delta x\to 0}\left(\frac{\Delta y}{\Delta x}\right)$$

$$= \lim_{\Delta x\to 0}\left(\frac{\sqrt{x+\Delta x} - \sqrt{x}}{\Delta x}\right).$$

This can be written as

$$\frac{\mathrm{d}y}{\mathrm{d}x} = \lim_{\Delta x\to 0}\left\{\frac{(\sqrt{x+\Delta x} - \sqrt{x})(\sqrt{x+\Delta x} + \sqrt{x})}{\Delta x\,(\sqrt{x+\Delta x} + \sqrt{x})}\right\},$$

i.e.

$$\frac{\mathrm{d}y}{\mathrm{d}x} = \lim_{\Delta x\to 0}\left\{\frac{(x+\Delta x) - x}{\Delta x\,(\sqrt{x+\Delta x} + \sqrt{x})}\right\}$$

$$= \lim_{\Delta x\to 0}\left(\frac{1}{\sqrt{x+\Delta x} + \sqrt{x}}\right)$$

$$= \frac{1}{2\sqrt{x}}.$$

Example 1.5

Find the gradient of the curve $y = \sqrt{x}$ at the point where $x = 9$. Hence find the equation of the straight line which is tangent to the curve $y = \sqrt{x}$ at the point (9, 3). Find also the equation of the *normal* to $y = \sqrt{x}$ at (9, 3). (The *normal* is the line passing through (9, 3) which is perpendicular to the tangent.)

The gradient of the curve $y = \sqrt{x}$ at a point is the value of $\mathrm{d}y/\mathrm{d}x$ at that point. From Example 1.4,

$$\frac{\mathrm{d}y}{\mathrm{d}x} = \frac{1}{2\sqrt{x}}.$$

Therefore, when $x = 9$, the gradient of $y = \sqrt{x}$ is $\frac{1}{6}$. The tangent to $y = \sqrt{x}$ at (9, 3) has a gradient $\frac{1}{6}$ and an equation $y - 3 = \frac{1}{6}(x - 9)$, i.e. the equation of the tangent is $x - 6y + 9 = 0$.

The normal to $y = \sqrt{x}$ at (9, 3) has a gradient -6 and an equation $y - 3 = -6(x - 9)$, i.e. the equation of the normal is $6x + y - 57 = 0$.

Example 1.6

The distance s m travelled by a moving particle is given by $s = \sqrt{t}$ where t is measured in seconds. Obtain an expression for the velocity v m/s in terms of t. Confirm that $v \to 0$ as $t \to \infty$.

$$s = \sqrt{t},$$

and, from Example 1.4,

$$\frac{\mathrm{d}s}{\mathrm{d}t} = \frac{1}{2\sqrt{t}}.$$

Therefore, $v = 1/2\sqrt{t}$ m/s and, clearly, $v \to 0$ as $t \to \infty$. It is important to understand that v *never reaches* the value zero but becomes arbitrarily close to zero. Note that $s \to \infty$ as $t \to \infty$.

Problems

1. Use the definition of a derivative to obtain $f'(x)$ for the following functions.
   (i) $f(x) = x^3$.
   (ii) $f(x) = 1/x$.

2. Show that the derivative of $kf(x)$ is $kf'(x)$, where k is a constant. Hence *write down* the derivative of the functions $4x^3$ and $1/3x$.

3. Use the definition of a derivative to obtain $f'(x)$ when $f(x) = 20x + 2x^2$. Is it true that $f'(x)$ = derivative of $20x$ + derivative of $2x^2$?

4. The distance travelled by a moving vehicle is given by $s = 20t + 2t^2$ where s is measured in kilometres and t in hours. Obtain an expression for the speed $\mathrm{d}s/\mathrm{d}t$ km/h of the vehicle after t h. What is the speed after the following times?
   (i) 30 min.
   (ii) 2 h.

5. Show that the function $y = |x|$ does not possess a derivative when $x = 0$. (We say that the function $|x|$ is **not differentiable** at $x = 0$.)

6. Find the equation of the tangent and the normal to the curve $y = x^2$ at the point $(2, 4)$.

Answers

1. (i) $3x^2$.
   (ii) $-\dfrac{1}{x^2}$.
2. $12x^2, \; -\dfrac{1}{3x^2}$.
3. $20 + 4x$.
4. $20 + 4t$ km/h.
   (i) 22 km/h.
   (ii) 28 km/h.
6. The equation of the tangent is $y = 4x - 4$; the equation of the normal is $4y = -x + 18$.

1.4 THE DERIVATIVES OF POLYNOMIALS, AND THE FUNCTION OF A FUNCTION RULE

In Example 1.3 (ii), we derived the rule for differentiating the function $y = x^n$ for the case when n is a positive integer, i.e. $y = x^n$ implies that $\mathrm{d}y/\mathrm{d}x = nx^{n-1}$. This rule can be written in compact form by elimination of y to give

$$\frac{\mathrm{d}}{\mathrm{d}x}(x^n) = nx^{n-1}. \tag{1.5}$$

Particular examples of equation (1.5) are $(\mathrm{d}/\mathrm{d}x)(x^2) = 2x$ (see Example 1.3 (i)), $(\mathrm{d}/\mathrm{d}x)(x^3) = 3x^2$ (see problem 1 (i)), and $(\mathrm{d}/\mathrm{d}x)(x) = 1$, (the gradient of the line $y = x$). Additionally, equation (1.5) is clearly true for the case $n = 0$, $(\mathrm{d}/\mathrm{d}x)(1) = 0$, the gradient of the line $y = 1$.

Further, the following results, obtained in section 1.3, suggest that the rule for differentiating x^n can be extended to fractional and negative values of n: $(\mathrm{d}/\mathrm{d}x)(x^{1/2}) = \frac{1}{2}x^{-1/2}$ (see Example 1.4) and $(\mathrm{d}/\mathrm{d}x)(x^{-1}) = -x^{-2}$ (see problem 1 (ii)). In fact, equation (1.5) is true for all values of n; the proofs for fractional and negative n are given in Chapter 7.

The rules for differentiating $cf(x)$, where c is a constant, and for the differentiation of the *sum* and *difference* of functions are easily obtained using the definition of a derivative together with Theorem 1.1 (i):

$$\frac{d}{dx}\{cf(x)\} = c\frac{d}{dx}\{f(x)\}, \tag{1.6}$$

$$\frac{d}{dx}\{f(x) \pm g(x)\} = \frac{d}{dx}\{f(x)\} \pm \frac{d}{dx}\{g(x)\}. \tag{1.7}$$

The special rules for differentiating *products* and *quotients* of functions are dealt with in Chapter 2.

Equations (1.5)–(1.7) may be applied to the differentiation of polynomials. For example, the quadratic polynomial $3x^2 + 4x - 1$ is differentiated term by term as follows:

$$\begin{aligned}\frac{d}{dx}(3x^2 + 4x - 1) &= \frac{d}{dx}(3x^2) + \frac{d}{dx}(4x) - \frac{d}{dx}(1)\\ &= 3\frac{d}{dx}(x^2) + 4\frac{d}{dx}(x) - \frac{d}{dx}(1)\\ &= 3(2x) + 4(1) - 0\\ &= 6x + 4.\end{aligned}$$

Clearly, it is unnecessary to write down the trivial intermediate steps.

The rule for differentiating x^n can be extended by applying Theorem 1.2.

Theorem 1.2 (The function of a function rule)
Given that y is a function of a variable u and that u is a function of x (i.e. y depends on x), then

$$\frac{dy}{dx} = \frac{dy}{du}\,\frac{du}{dx}.$$

Proof
Let Δx be a change in the variable x which results in changes Δu in u and Δy in y. On the assumption that $u(x)$ and $y(u)$ are differentiable functions, then Δu and Δy both tend to zero as $\Delta x \to 0$.

$$\begin{aligned}\frac{dy}{dx} &= \lim_{\Delta x \to 0}\left(\frac{\Delta y}{\Delta x}\right)\\ &= \lim_{\Delta x \to 0}\left(\frac{\Delta y}{\Delta u}\,\frac{\Delta u}{\Delta x}\right)\end{aligned}$$

By Theorem 1.1 (ii),

$$\lim_{\Delta x\to 0}\left(\frac{\Delta y}{\Delta u}\,\frac{\Delta u}{\Delta x}\right) = \lim_{\Delta x\to 0}\left(\frac{\Delta y}{\Delta u}\right)\lim_{\Delta x\to 0}\left(\frac{\Delta u}{\Delta x}\right).$$

$$= \lim_{\Delta u\to 0}\left(\frac{\Delta y}{\Delta u}\right)\lim_{\Delta x\to 0}\left(\frac{\Delta u}{\Delta x}\right),$$

and, by definition of a derivative, this equals

$$\frac{dy}{du}\,\frac{du}{dx}.$$

This powerful result can be written in the alternative form

$$\frac{d}{dx}\{f(u)\} = f'(u)\,\frac{du}{dx} \tag{1.8}$$

and can be applied to extend *any* differentiation rule. In particular, equation (1.5) is extended to give

$$\frac{d}{dx}(u^n) = nu^{n-1}\,\frac{du}{dx} \tag{1.9a}$$

or, alternatively, if u is the function $u = g(x)$,

$$\frac{d}{dx}\{g(x)\}^n = n\{g(x)\}^{n-1}g'(x). \tag{1.9b}$$

Differentiation using equations (1.9) is demonstrated in Examples 1.8 and 1.9 and further applications of Theorem 1.2 are considered in Chapter 7.

Example 1.7

(i) Determine dy/dx when $y = 16x^5 - 3x^4 + 4x^3 - 3x + 7$.

(ii) Determine $f'(t)$ when

$$f(t) = \sqrt{t} - \frac{7}{\sqrt{t}} + \frac{9}{t^2}.$$

(i) $y = 16x^5 - 3x^4 + 4x^3 - 3x + 7$.

$$\frac{dy}{dx} = 16(5x^4) - 3(4x^3) + 4(3x^2) - 3 + 0$$

$$= 80x^4 - 12x^3 + 12x^2 - 3.$$

(ii) $f(t) = t^{1/2} - 7t^{-1/2} + 9t^{-2}$.

$$f'(t) = \tfrac{1}{2}t^{-1/2} - 7(-\tfrac{1}{2}t^{-3/2}) + 9(-2t^{-3})$$

$$= \frac{1}{2\sqrt{t}} + \frac{7}{2\sqrt{t^3}} - \frac{18}{t^3}.$$

Example 1.8

Differentiate with respect to x the function $f(x) = (x^2 + 1)^3$ by the following methods.

(i) Expanding the binomial.

(ii) Using the function of a function rule.

(i) Let $y = (x^2 + 1)^3$. Expanding, we obtain

$$y = x^6 + 3x^4 + 3x^2 + 1.$$

$$\frac{dy}{dx} = 6x^5 + 12x^3 + 6x$$

$$= 6x(x^2 + 1)^2.$$

(ii) Let $y = u^3$, where $u = x^2 + 1$.

$$\frac{dy}{dx} = \frac{dy}{du}\frac{du}{dx}$$

$$= (3u^2)(2x)$$

$$= 3(x^2 + 1)^2(2x)$$

$$= 6x(x^2 + 1)^2.$$

Example 1.9

Find $f'(x)$ for the following functions.

(i) $f(x) = (x^2 + 1)^{10}$.

(ii) $f(x) = (x^3 - 3x)^4$.

(iii) $f(x) = \sqrt{x + \frac{1}{x}}$.

(i) Let $y = u^{10}$, where $u = x^2 + 1$.

$$\frac{dy}{dx} = \frac{dy}{du}\frac{du}{dx}$$

$$= (10u^9)(2x).$$

Therefore, $f'(x) = 20x(x^2 + 1)^9$. With practice you should be able to do differentiations of this type without introducing the variable u, i.e. by applying equation (1.9b) directly. (ii) and (iii) are completed using equation (1.9b).

(ii) $f(x) = (x^3 - 3x)^4$.

$$f'(x) = 4(x^3 - 3x)^3 \frac{d}{dx}(x^3 - 3x)$$

$$= 4(x^3 - 3x)^3(3x^2 - 3)$$

$$= 12(x^3 - 3x)^3(x^2 - 1).$$

(iii) $f(x) = \sqrt{x + \frac{1}{x}}$

$$= (x + x^{-1})^{1/2}.$$

$$f'(x) = \frac{1}{2}(x + x^{-1})^{-1/2} \frac{d}{dx}(x + x^{-1})$$

$$= \frac{1}{2}(x + x^{-1})^{-1/2}(1 - x^{-2})$$

$$= \frac{1}{2}\left(x + \frac{1}{x}\right)^{-1/2}\left(1 - \frac{1}{x^2}\right).$$

Example 1.10
Given that p is a function of v defined by the equation

$$p = \frac{3}{1 + 2v^2},$$

determine the rate of change in p with respect to v when $v = 1$.

We require the value of dp/dv when $v = 1$. Let $p = 3u^{-1}$ where $u = 1 + 2v^2$.

$$\frac{dp}{dv} = \frac{dp}{du}\frac{du}{dv}$$

$$= (-3u^{-2})(4v)$$

$$= -\frac{12v}{(1 + 2v^2)^2}$$

$$= -\tfrac{4}{3} \quad \text{when} \quad v = 1.$$

Example 1.11
A large spherical balloon is being inflated such that its volume is increasing at a constant rate of 2 m^3/min. Calculate the rate of increase in the radius of the ballon at the instant when the radius is 1 m.

Let the radius and volume of the balloon be denoted by r and V respectively. We are *given* that $\mathrm{d}V/\mathrm{d}t = 2\ \mathrm{m}^3/\mathrm{min}$, and we require to obtain $\mathrm{d}r/\mathrm{d}t$. V and r are related by the function $V = \frac{4}{3}\pi r^3$ and, applying the function of a function rule, we have

$$\frac{\mathrm{d}V}{\mathrm{d}t} = \frac{\mathrm{d}V}{\mathrm{d}r}\,\frac{\mathrm{d}r}{\mathrm{d}t}$$

$$= 4\pi r^2\,\frac{\mathrm{d}r}{\mathrm{d}t}.$$

Therefore, $2 = 4\pi r^2\ (\mathrm{d}r/\mathrm{d}t)$ and, at the instant when $r = 1$, $\mathrm{d}r/\mathrm{d}t = 2/4\pi \approx 0.159$ m/min.

Problems of this type, known as *related rates* problems, are dealt with in more detail in Chapter 7.

Problems

1. Differentiate the following functions with respect to x.
   (i) $7x^3 + 4x^2 - 3x + 18$.
   (ii) $(x^2 + 1)(x + 2)$.
   (iii) $\sqrt{x} + \sqrt[3]{x}$.
   (iv) $2x^{3/2} - x^{-1/4} + 24x$.
   (v) $(x^2 + 2)^2$.
   (vi) $(x^2 + 2)^{12}$.

2. (i) Find $\mathrm{d}p/\mathrm{d}v$ when

$$p = \frac{2}{\sqrt{v}} - \frac{3}{\sqrt[3]{v}} + 14.$$

   (ii) Find $\mathrm{d}s/\mathrm{d}t$ when

$$s = t^4 - t^2 + 1 - t^{-2} + t^{-4}.$$

   (iii) Find $\mathrm{d}F/\mathrm{d}r$ when

$$F = \frac{12}{(7 + r)^2}.$$

3. Determine the following.

   (i) $$\frac{\mathrm{d}}{\mathrm{d}x}\{(4x^3 + 7x + 3)^6\}.$$

   (ii) $$\frac{\mathrm{d}}{\mathrm{d}x}\left\{\frac{1}{(1 - x)^2}\right\}.$$

   (iii) $$\frac{\mathrm{d}}{\mathrm{d}x}\{\sqrt{3x^2 + 4x - 3}\}.$$

(iv) $\dfrac{\mathrm{d}}{\mathrm{d}x}\{(x^2 + 2x + 1)^{5/2}\}$.

(v) $\dfrac{\mathrm{d}}{\mathrm{d}t}\{(3 - t - 4t^2)^8\}$.

(vi) $\dfrac{\mathrm{d}}{\mathrm{d}s}\left(\sqrt{1 + \sqrt{s}}\right)$.

4. Find the equations of the tangent and the normal to the curve

$$y = \frac{9}{1 + 2x}$$

at the point $(1, 3)$.

5. A body is thrown vertically upwards with a velocity of 20 m/s. If the distance s m above the ground after t s is given by $s = 20t - 5t^2$, find the following.
 (i) The velocity after t s.
 (ii) The highest point to which it will rise.
 (iii) When it will strike the ground.

6. The profit £P made by a factory when it produces an x kg batch of a certain commodity is given by

$$P = 15x^2 - 60 - \frac{x^3}{3}.$$

Find the rate at which the profit changes with respect to the number of kilograms produced for the following x values.
 (i) $x = 10$ kg.
 (ii) $x = 20$ kg.
 (iii) $x = 35$ kg.
For what value of x is the rate of change in profit equal to zero?

7. The force F N between two charged particles at a distance r m apart is given by

$$F = \frac{c}{r^2}$$

where c is a constant. Determine an expression for the rate of decrease in force with respect to increasing distance.

8. At a certain time the number of bacteria in a culture was measured to be 5×10^4. Experimentation confirmed that the number present t hours later was given by the formula

$$N = 5 \times 10^4 (1 + t)^2.$$

Determine the rate of growth of the bacteria when $t = 3$ hours.

9. When temperature is constant the pressure p and the volume v of a gas are related by the law

$$pv = \text{constant}.$$

Show that

$$\frac{dp}{dv} = -\frac{p}{v}.$$

Answers

1. (i) $21x^2 + 8x - 3$.
(ii) $3x^2 + 4x + 1$.
(iii) $\dfrac{1}{2\sqrt{x}} + \dfrac{1}{3\sqrt[3]{x^2}}$.
(iv) $3x^{1/2} + \frac{1}{4}x^{-5/4} + 24$.
(v) $4x(x^2 + 2)$.
(vi) $24x(x^2 + 2)^{11}$.

2. (i) $-\dfrac{1}{\sqrt{v^3}} + \dfrac{1}{\sqrt[3]{v^4}}$.
(ii) $4t^3 - 2t + 2t^{-3} - 4t^{-5}$.
(iii) $-\dfrac{24}{(7 + r)^3}$.

3. (i) $6(12x^2 + 7)(4x^3 + 7x + 3)^5$.
(ii) $\dfrac{2}{(1 - x)^3}$.
(iii) $\dfrac{3x + 2}{\sqrt{3x^2 + 4x - 3}}$.
(iv) $5(x + 1)^4$.
(v) $-8(1 + 8t)(3 - t - 4t^2)^7$.
(vi) $\frac{1}{4}(s\sqrt{s} + s)^{-1/2}$.

4. The equation of the tangent is $2x + y - 5 = 0$; the equation of the normal is $x - 2y + 5 = 0$.

5. (i) $20 - 10t$ m/s.
(ii) 20 m.
(iii) After 4 s.

6. (i) £200 per kilogram.
   (ii) £200 per kilogram.
   (iii) −£175 per kilogram.
   $\mathrm{d}P/\mathrm{d}x = 0$ when $x = 30$ kg.
7. F decreases at a rate $2c/r^3$ N/m as the distance increases.
8. 4×10^5 bacteria per hour.

1.5 THE DERIVATIVES OF SIN X AND COS X

The definition of the derivative of a function, equation (1.2), may be applied to the function $f(x) = \sin x$ as follows:

$$f'(x) = \lim_{h \to 0} \left\{ \frac{\sin (x+h) - \sin x}{h} \right\}.$$

Using equation (0.14),

$$f'(x) = \lim_{h \to 0} \left\{ \frac{2 \cos (x + h/2) \sin (h/2)}{h} \right\}.$$

From equation (1.1), the small-angle approximation

$$f'(x) = \lim_{h \to 0} \left[\frac{2 \{\cos (x + h/2)\} (h/2)}{h} \right]$$

$$= \lim_{h \to 0} \left\{ \cos \left(x + \frac{h}{2} \right) \right\}$$

$$= \cos x.$$

Therefore, we have established the rule

$$\frac{\mathrm{d}}{\mathrm{d}x} (\sin x) = \cos x. \tag{1.10}$$

A similar proof leads to the rule

$$\frac{\mathrm{d}}{\mathrm{d}x} (\cos x) = -\sin x. \tag{1.11}$$

Alternatively, equation (1.11) can be derived using the function of a function rule as follows:

$$y = \cos x = \sin \left(x + \frac{\pi}{2} \right) = \sin u,$$

where

$$u = x + \frac{\pi}{2}.$$

Now

$$\frac{dy}{dx} = \frac{dy}{du}\,\frac{du}{dx}$$

and, using equation (1.10), we have

$$\begin{aligned}\frac{dy}{du}\,\frac{du}{dx} &= (\cos u)\,(1)\\ &= \cos\left(x + \frac{\pi}{2}\right)\\ &= -\sin x.\end{aligned}$$

The extended standard forms for the functions $\sin x$ and $\cos x$ are obtained by applying equation (1.8):

$$\frac{d}{dx}(\sin u) = \cos u\,\frac{du}{dx} \tag{1.12a}$$

$$\frac{d}{dx}(\cos u) = -\sin u\,\frac{du}{dx} \tag{1.13a}$$

or, alternatively,

$$\frac{d}{dx}\left[\sin\{g(x)\}\right] = \cos\{g(x)\}\,g'(x) \tag{1.12b}$$

$$\frac{d}{dx}\left[\cos\{g(x)\}\right] = -\sin\{g(x)\}\,g'(x). \tag{1.13b}$$

Example 1.12

Differentiate with respect to x the following functions.

(i) $\sin 2x$.

(ii) $\cos\left(\frac{x}{2}\right)$.

(iii) $\cos(1 - 4x)$.

(i) Let $y = \sin 2x = \sin u$ where $u = 2x$.

$$\frac{dy}{dx} = \frac{dy}{du}\frac{du}{dx}$$

$$= (\cos u)(2)$$

$$= 2 \cos 2x.$$

(ii) Let

$$y = \cos\left(\frac{x}{2}\right) = \cos u$$

where

$$u = \frac{x}{2}.$$

$$\frac{dy}{dx} = \frac{dy}{du}\frac{du}{dx}$$

$$= (-\sin u)\left(\frac{1}{2}\right)$$

$$= -\frac{1}{2}\sin\left(\frac{x}{2}\right).$$

As we saw in Example 1.9 it is not necessary to introduce the variable u. (iii) is completed directly using equation (1.13b).

(iii) $$\frac{d}{dx}\{\cos(1-4x)\} = -\sin(1-4x)\frac{d}{dx}(1-4x)$$

$$= -\sin(1-4x)(-4)$$

$$= 4\sin(1-4x).$$

Example 1.13

Determine the following.

(i) $\dfrac{d}{dx}\{\sin(x^2)\}$.

(ii) $\dfrac{d}{dt}(\sin^2 t)$.

(iii) $\dfrac{d}{dx}(\cos^3 2x)$.

(i) $$\frac{d}{dx}\{\sin(x^2)\} = \cos(x^2)\frac{d}{dx}(x^2)$$
$$= 2x\cos(x^2).$$

(ii) $\sin^2 t$ is another way of writing $(\sin t)^2$. Therefore, from equation (1.9b),

$$\frac{d}{dt}\{(\sin t)^2\} = 2\sin t\frac{d}{dt}(\sin t)$$
$$= 2\sin t\cos t.$$

(iii) $$\frac{d}{dx}(\cos^3 2x) = \frac{d}{dx}\{(\cos 2x)^3\}.$$

From equation (1.9b),

$$\frac{d}{dx}\{(\cos 2x)^3\} = 3(\cos 2x)^2\frac{d}{dx}(\cos 2x),$$

and, from equation (1.13b),

$$3(\cos 2x)^2\frac{d}{dx}(\cos 2x) = 3(\cos 2x)^2(-\sin 2x)\frac{d}{dx}(2x)$$
$$= 3(\cos 2x)^2(-\sin 2x)(2)$$
$$= -6\cos^2 2x\sin 2x.$$

Example 1.14

Given that

$$y = \frac{1}{\sin x},$$

show that

$$\frac{dy}{dx} = -\frac{\cos x}{\sin^2 x}.$$

$$y = (\sin x)^{-1}.$$

From equation (1.9b),

$$\frac{dy}{dx} = -(\sin x)^{-2}\frac{d}{dx}(\sin x)$$
$$= -(\sin x)^{-2}\cos x$$
$$= -\frac{\cos x}{\sin^2 x}.$$

Problems

1. Differentiate with respect to x the following functions.

   (i) $\cos 3x$.

   (ii) $2 \sin \left(\dfrac{\pi}{3} - \dfrac{x}{2}\right)$.

   (iii) $\sin (x^3 + 3x)$.

   (iv) $\sqrt{\sin 2x}$.

   (v) $\sin^3 x$.

   (vi) $\cos^2 \left(\dfrac{x}{2}\right)$.

2. Determine

   (i) $\dfrac{\mathrm{d}}{\mathrm{d}\theta} (\cos \theta - \tfrac{1}{3} \cos^3 \theta)$.

   (ii) $\dfrac{\mathrm{d}}{\mathrm{d}t} [4t - \sin^2 \{2(1-t)\}]$.

   (iii) $\dfrac{\mathrm{d}}{\mathrm{d}x}\left(\dfrac{1}{\cos x}\right)$.

3. Show that the curve $y = 3t + \sin 2t$ has a positive gradient for all values of t. Find the equation of the tangent to the curve at the point when $t = \pi/2$.

4. The position s cm of an oscillating component in a mechanical device is given at time t s by the equation

   $$s = R \sin (\omega t + \phi),$$

   where R, ω and ϕ are constants. Show that the acceleration a cm/s^2 satisfies the relationship $a = -\omega^2 s$.

Answers

1. (i) $-3 \sin 3x$.

   (ii) $-\cos\left(\dfrac{\pi}{3} - \dfrac{x}{2}\right)$.

   (iii) $(3x^2 + 3) \cos (x^3 + 3x)$.

   (iv) $\dfrac{\cos 2x}{\sqrt{\sin 2x}}$.

(v) $3 \sin^2 x \cos x$.

(vi) $-\cos\left(\frac{x}{2}\right) \sin\left(\frac{x}{2}\right)$.

2. (i) $-\sin\theta + \cos^2\theta \sin\theta$.
(ii) $4 + 4\sin\{2(1-t)\}\cos\{2(1-t)\}$.

(iii) $\dfrac{\sin x}{\cos^2 x}$.

3. $y = t + \pi$.

Further problems for Chapter 1

('Further problems' will appear at the end of each chapter to test the understanding of the fundamental material contained in the chapter. Each problem is *short* and should take *less than 2 min*. Readers should be reluctant to continue to the next chapter until they have been successful with the majority of these problems.)

1. Given that

$$f(x) = \frac{x^2 - x - 6}{2x^2 - 6x},$$

evaluate the following.

(i) $\lim_{x \to 3} \{f(x)\}$.

(ii) $\lim_{x \to \infty} \{f(x)\}$.

2. Evaluate

$$\lim_{x \to 0} \left(\frac{x^2 + 2x}{\sin x}\right).$$

3. Write down the value of

$$\lim_{h \to 0} \left\{\frac{(x+h)^4 - x^4}{h}\right\}$$

when $x = 2$.

4. Determine the gradient of the curve $y = 2x^3 - 3x + 1$ at the point $(1, 0)$.

5. The position s of a moving particle at time t is given by $s = 7t - 5t^2$. Show that the acceleration is constant.

6. Determine the equation of the tangent to the curve $y = 7 - x^2$ at the point (2, 3).

7. Given that $f(x) = (2x^2 - 3)^4$, calculate $f'(-1)$.

8. Differentiate with respect to t the function

$$f(t) = \left(3\sqrt{t} - \frac{1}{t}\right)^2.$$

9. Write down the value of

$$\lim_{h \to 0}\left\{\frac{\cos(x + h) - \cos x}{h}\right\},$$

when $x = \pi/2$.

10. Evaluate the gradient of the curve

$$y = \sin\left(\frac{\pi}{3} - 7x\right)$$

when $x = 0$.

11. Evaluate

$$\frac{d}{d\theta}\left\{\cos\left(\frac{\pi}{2} - 2\pi\theta^2\right)\right\}$$

when $\theta = 1$.

12. Determine

$$\frac{d}{ds}(\cos^2 s).$$

13. The position s m at time t s of a moving component is given by $s = 0.7 \sin 30t$. Determine the maximum value of the velocity of the component.

14. Given that $y = 3x^2$ and $dx/dt = 4$ evaluate dy/dt when $x = 1$.

Answers

1. (i) $\frac{5}{6}$.
   (ii) $\frac{1}{2}$.
2. 2.
3. 32.
4. 3.

6. $y + 4x - 11 = 0$.
7. 16.
8. $9 + 3t^{-3/2} - 2t^{-3}$.
9. -1.
10. -3.5,
11. 4π.
12. $-2 \cos s \sin s$.
13. 21 m/s.
14. 24.

2

Differentiation of Products and Quotients

2.1 INTRODUCTION

In Chapter 1 we saw that the derivative of the sum (or difference) of two functions was simply the sum (or difference) of their derivatives. However, the derivative of the product (or quotient) of two functions is *not* the product (or quotient) of the derivatives. The *special rules* which apply when differentiating products and quotients are given in Theorem 2.1.

The rule for differentiating quotients may be applied to the functions

$$\tan x = \frac{\sin x}{\cos x}$$

and

$$\cot x = \frac{\cos x}{\sin x}$$

and in section 2.3 a summary is given of the differentiation rules for the six trigonometric functions.

2.2 THE PRODUCT RULE AND THE QUOTIENT RULE

Suppose that $u = f(x)$ and $v = g(x)$ are differentiable functions of x. The rules for differentiating the *product* uv and the *quotient* u/v are stated in Theorem 2.1.

Theorem 2.1

$$\frac{\mathrm{d}}{\mathrm{d}x}(uv) = u\,\frac{\mathrm{d}v}{\mathrm{d}x} + v\frac{\mathrm{d}u}{\mathrm{d}x}. \tag{2.1a}$$

$$\frac{\mathrm{d}}{\mathrm{d}x}\left(\frac{u}{v}\right) = \frac{v(\mathrm{d}u/\mathrm{d}x) - u(\mathrm{d}v/\mathrm{d}x)}{v^2}. \tag{2.2a}$$

Proof
Let Δx be a small change in x resulting in changes Δu in u, Δv in v and Δy in y where $y = uv$. Therefore,

$$y + \Delta y = (u + \Delta u)(v + \Delta v)$$
$$= uv + u\,\Delta v + v\,\Delta u + \Delta u\,\Delta v.$$

Thus,

$$\Delta y = u\,\Delta v + v\,\Delta u + \Delta u\,\Delta v$$

and

$$\frac{\Delta y}{\Delta x} = u\,\frac{\Delta v}{\Delta x} + v\frac{\Delta u}{\Delta x} + \Delta u\,\frac{\Delta v}{\Delta x}.$$

Now let $\Delta x \to 0$ (also $\Delta u \to 0$, since u is differentiable), and by the definition of a derivative we obtain

$$\frac{\mathrm{d}y}{\mathrm{d}x} = u\,\frac{\mathrm{d}v}{\mathrm{d}x} + v\,\frac{\mathrm{d}u}{\mathrm{d}x}.$$

A similar proof may be used to prove equation (2.2a), but it is simpler to write $y = u/v$ and to apply equation (2.1a) to the product $u = yv$ to give

$$\frac{\mathrm{d}u}{\mathrm{d}x} = y\,\frac{\mathrm{d}v}{\mathrm{d}x} + v\,\frac{\mathrm{d}y}{\mathrm{d}x}.$$

Therefore,

$$\frac{\mathrm{d}y}{\mathrm{d}x} = \frac{1}{v}\left(\frac{\mathrm{d}u}{\mathrm{d}x} - \frac{u}{v}\frac{\mathrm{d}v}{\mathrm{d}x}\right)$$

$$= \frac{v(\mathrm{d}u/\mathrm{d}x) - u(\mathrm{d}v/\mathrm{d}x)}{v^2}.$$

The *product rule* and *quotient rule* may be written in functional notation as follows:

$$\frac{\mathrm{d}}{\mathrm{d}x}\{f(x)g(x)\} = f(x)\,g'(x) + g(x)\,f'(x), \tag{2.1b}$$

$$\frac{\mathrm{d}}{\mathrm{d}x}\left\{\frac{f(x)}{g(x)}\right\} = \frac{g(x)f'(x) - f(x)g'(x)}{\{g(x)\}^2} \tag{2.2b}$$

Example 2.1
Differentiate the following with respect to x.
(i) $(x^2 + 4)(x^3 + 3x^2 + 7x + 1)$.
(ii) $x^2 \sin 2x$.

(i) Let $y = (x^2 + 4)(x^3 + 3x^2 + 7x + 1)$. The differentiation can be done by multiplying the two polynomials before attempting to differentiate. This can be avoided by applying the product rule, equation (2.1a), with $u = x^2 + 4$ and $v = x^3 + 3x^2 + 7x + 1$. Thus,

$$\frac{\mathrm{d}y}{\mathrm{d}x} = (x^2 + 4)(3x^2 + 6x + 7) + (x^3 + 3x^2 + 7x + 1)(2x).$$

(ii) It is not necessary to introduce the functions u and v if the product rule is used in the form of equation (2.1b).

$$\frac{\mathrm{d}}{\mathrm{d}x}(x^2 \sin 2x) = x^2 \frac{\mathrm{d}}{\mathrm{d}x}(\sin 2x) + \sin 2x \frac{\mathrm{d}}{\mathrm{d}x}(x^2)$$
$$= 2x^2 \cos 2x + 2x \sin 2x.$$

Example 2.2
Use the quotient rule to obtain $\mathrm{d}y/\mathrm{d}x$ for the following.

(i) $y = \dfrac{1 + x^2}{1 - x^2}$.

(ii) $y = \dfrac{x}{\sqrt{1 - x^2}}$.

In (i), the quotient rule as given in equation (2.2a) is demonstrated while, in (ii), equation (2.2b) is applied.
(i) Let $u = 1 + x^2, v = 1 - x^2$. Therefore,

$$\frac{\mathrm{d}u}{\mathrm{d}x} = 2x,$$

$$\frac{\mathrm{d}v}{\mathrm{d}x} = -2x$$

and

$$\frac{dy}{dx} = \frac{(1-x^2)\times 2x - (1+x^2)(-2x)}{(1-x^2)^2}$$

$$= \frac{4x}{(1-x^2)^2}.$$

(ii) $$\frac{dy}{dx} = \frac{d}{dx}\left\{\frac{x}{(1-x^2)^{1/2}}\right\}$$

$$= \frac{(1-x^2)^{1/2}(d/dx)(x) - x(d/dx)\{(1-x^2)^{1/2}\}}{\{(1-x^2)^{1/2}\}^2}$$

$$= \frac{(1-x^2)^{1/2}\times 1 - x\,\frac{1}{2}(1-x^2)^{-1/2}(-2x)}{1-x^2}$$

$$= \frac{1-x^2+x^2}{(1-x^2)^{3/2}}$$

$$= \frac{1}{(1-x^2)^{3/2}}.$$

(Note that, when differentiating quotients, it is usually *not* a good idea to convert them to products by writing, for example,

$$\frac{x}{\sqrt{1-x^2}} = x(1-x^2)^{-1/2}.$$

This procedure can make simplification of the result more difficult.)

Problems

1. Differentiate the following with respect to x.
   (i) $(x^2 + 2x)\cos x$.
   (ii) $x^2\sqrt{1+x^2}$.
   (iii) $(1+x^2)^2 \cos 3x$.
   (iv) $\dfrac{x^2+1}{x^2-3}$.
   (v) $\left(\dfrac{1-x}{1+x}\right)^2$.
   (vi) $\dfrac{1+\sin 2x}{1+\cos 2x}$.

2. Determine the following.

(i) $\dfrac{d}{dt}\{(t^2 + 2t - 1)(3t^2 - 5t + 2)\}$.

(ii) $\dfrac{d}{dx}\left(\dfrac{x^2}{x - 1}\right)$.

(iii) $\dfrac{d}{dz}\left(\dfrac{z^2 + z - 1}{z^2 + z + 1}\right)$.

(iv) $\dfrac{d}{du}\left(\dfrac{u + 3}{u - 3}\right)$.

(v) $\dfrac{d}{dx}\{(4 - x)\sqrt{4 + x}\}$.

(vi) $\dfrac{d}{dx}\left(\sqrt{\dfrac{1 - x^2}{1 + x^2}}\right)$.

3. Obtain dy/dx for the following.

(i) $y = \dfrac{\sin 2x}{\sin x + \cos x}$.

(ii) $y = x^2 \sin x + \sqrt{\cos x}$.

(iii) $y = \sin^3 x \cos x$.

(iv) $y = \dfrac{\sin^2 x}{x^2}$.

(v) $y = (x^2 + 1)^3 \cos^4 2x$.

4. The distance of a moving component from its initial position is s m, given by

$$s = 0.04t \sin 100\pi t,$$

where time t is measured in seconds.

(i) What is the velocity of the component when $t = 3/400$?

(ii) Show that the acceleration a m/s^2 satisfies the equation

$$a + (100\pi)^2 s = 8\pi \cos 100\pi t.$$

Answers

1. (i) $(2x + 2)\cos x - (x^2 + 2x)\sin x$.

(ii) $\dfrac{x^3}{\sqrt{1 + x^2}} + 2x\sqrt{1 + x^2}$.

(iii) $4x(1+x^2)\cos 3x - 3(1+x^2)^2 \sin 3x.$

(iv) $-\left\{\dfrac{x^4+3x^2+6x}{(x^2-3)^2}\right\}.$

(v) $-\dfrac{4(1-x)}{(1+x)^3}.$

(vi) $\dfrac{2(1+\sin 2x+\cos 2x)}{(1+\cos 2x)^2}.$

2. (i) $12t^3+3t^2-22t+9.$

(ii) $\dfrac{x^2-2x}{(x-1)^2}.$

(iii) $\dfrac{4z+2}{(z^2+z+1)^2}.$

(iv) $-\dfrac{6}{(u-3)^2}.$

(v) $-\dfrac{4+3x}{2\sqrt{4+x}}.$

(vi) $-\dfrac{2x}{(1+x^2)^{3/2}(1-x^2)^{1/2}}.$

3. (i) $\dfrac{2(\cos^3 x-\sin^3 x)}{(\cos x+\sin x)^2}.$

(ii) $2x\sin x + x^2\cos x - \dfrac{\sin x}{2\sqrt{\cos x}}.$

(iii) $3\sin^2 x - 4\sin^4 x.$

(iv) $\dfrac{2\sin x}{x^3}(x\cos x-\sin x).$

(v) $2(x^2+1)^2\cos^3 2x\,\{3x\cos 2x - 4(x^2+1)\sin 2x\}.$

4. (i) $\dfrac{1}{100\sqrt{2}}(4-3\pi)$ m/s.

2.3 THE DERIVATIVES OF THE SIX TRIGONOMETRIC FUNCTIONS

The functions $\sin x$, $\cos x$ and

$$\tan x = \frac{\sin x}{\cos x}$$

are well known; the remaining three trigonometric functions are defined as follows:

$$\operatorname{cosec} x = \frac{1}{\sin x},$$

$$\sec x = \frac{1}{\cos x},$$

$$\cot x = \frac{1}{\tan x} = \frac{\cos x}{\sin x}.$$

The derivatives of $\sin x$ and $\cos x$ were obtained in section 1.5. The derivatives of the other four trigonometric functions may be obtained by application of the quotient rule (see Example 2.3). The results are summarized below in extended form according to the style of equation (1.8) (u is a function of x in each case):

$$\frac{\mathrm{d}}{\mathrm{d}x}(\sin u) = \cos u \frac{\mathrm{d}u}{\mathrm{d}x}, \tag{2.3}$$

$$\frac{\mathrm{d}}{\mathrm{d}x}(\cos u) = -\sin u \frac{\mathrm{d}u}{\mathrm{d}x}, \tag{2.4}$$

$$\frac{\mathrm{d}}{\mathrm{d}x}(\tan u) = \sec^2 u \frac{\mathrm{d}u}{\mathrm{d}x}, \tag{2.5}$$

$$\frac{\mathrm{d}}{\mathrm{d}x}(\operatorname{cosec} u) = -\operatorname{cosec} u \cot u \frac{\mathrm{d}u}{\mathrm{d}x}, \tag{2.6}$$

$$\frac{\mathrm{d}}{\mathrm{d}x}(\sec u) = \sec u \tan u \frac{\mathrm{d}u}{\mathrm{d}x}, \tag{2.7}$$

$$\frac{\mathrm{d}}{\mathrm{d}x}(\cot u) = -\operatorname{cosec}^2 u \frac{\mathrm{d}u}{\mathrm{d}x}. \tag{2.8}$$

Example 2.3

Use the quotient rule to differentiate the functions $\tan x$ and $\operatorname{cosec} x$.

$$\begin{aligned}
\frac{\mathrm{d}}{\mathrm{d}x}(\tan x) &= \frac{\mathrm{d}}{\mathrm{d}x}\left(\frac{\sin x}{\cos x}\right) \\
&= \frac{\cos x\,(\mathrm{d}/\mathrm{d}x)(\sin x) - \sin x\,(\mathrm{d}/\mathrm{d}x)(\cos x)}{\cos^2 x} \\
&= \frac{\cos x \cos x - \sin x\,(-\sin x)}{\cos^2 x} \\
&= \frac{\cos^2 x + \sin^2 x}{\cos^2 x} \\
&= \frac{1}{\cos^2 x} \\
&= \sec^2 x.
\end{aligned}$$

Equation (2.5) follows immediately from this result.

$$\begin{aligned}
\frac{\mathrm{d}}{\mathrm{d}x}(\operatorname{cosec} x) &= \frac{\mathrm{d}}{\mathrm{d}x}\left(\frac{1}{\sin x}\right) \\
&= \frac{\sin x\,(\mathrm{d}/\mathrm{d}x)(1) - 1 \times (\mathrm{d}/\mathrm{d}x)(\sin x)}{\sin^2 x} \\
&= \frac{0 - \cos x}{\sin^2 x} \\
&= -\frac{1}{\sin x}\,\frac{\cos x}{\sin x} \\
&= -\operatorname{cosec} x \cot x.
\end{aligned}$$

Equation (2.6) follows immediately from this result which may also be obtained using the function of a function rule as follows:

$$\begin{aligned}
\frac{\mathrm{d}}{\mathrm{d}x}(\operatorname{cosec} x) &= \frac{\mathrm{d}}{\mathrm{d}x}\{(\sin x)^{-1}\} \\
&= -(\sin x)^{-2}\,\frac{\mathrm{d}}{\mathrm{d}x}(\sin x) \\
&= -(\sin x)^{-2} \cos x \\
&= -\operatorname{cosec} x \cot x.
\end{aligned}$$

Example 2.4

Differentiate the following with respect to x.

(i) $\sec 3x$.

(ii) $\sec^2 3x$.
(iii) $(x + 1)^2 \tan 5x$.

(i) Replacing u by $3x$ in equation (2.7), we obtain

$$\frac{d}{dx}(\sec 3x) = \sec 3x \tan 3x \frac{d}{dx}(3x)$$

$$= 3 \sec 3x \tan 3x.$$

(ii) Let $y = u^2$ where $u = \sec 3x$.

$$\frac{d}{dx}(\sec^2 3x) = \frac{dy}{dx}$$

$$= \frac{dy}{du}\frac{du}{dx}$$

From (i),

$$\frac{dy}{du}\frac{du}{dx} = (2u)(3 \sec 3x \tan 3x)$$

$$= 6 \sec^2 3x \tan 3x.$$

(iii) $$\frac{d}{dx}\{(x + 1)^2 \tan 5x\} = (x + 1)^2 \frac{d}{dx}(\tan 5x) + \tan 5x \frac{d}{dx}\{(x + 1)^2\}.$$

Now, from equation (2.5),

$$\frac{d}{dx}(\tan 5x) = \sec^2 5x \frac{d}{dx}(5x)$$

$$= 5 \sec^2 5x.$$

Therefore,

$$\frac{d}{dx}\{(x + 1)^2 \tan 5x\} = 5(x + 1)^2 \sec^2 5x + 2(x + 1) \tan 5x.$$

Example 2.5
Determine the equation of the tangent to the curve

$$y = \tfrac{1}{3} \tan^3 x + \tan x$$

at the point $(\pi/4, \tfrac{4}{3})$.

Let $y = \tfrac{1}{3}u^3 + u$, where $u = \tan x$

$$\begin{aligned}\frac{dy}{dx} &= \frac{dy}{du}\frac{du}{dx}\\ &= (u^2 + 1)\sec^2 x\\ &= (\tan^2 x + 1)\sec^2 x\\ &= \sec^4 x.\end{aligned}$$

The gradient of the curve when $x = \pi/4$ is the value of dy/dx at this point.

Therefore the gradient $= \sec^4(\pi/4) = (\sqrt{2})^4 = 4$, and the equation of the tangent is

$$y - \frac{4}{3} = 4\left(x - \frac{\pi}{4}\right),$$

i.e.

$$12x - 3y + 4 - 3\pi = 0.$$

Problems

1. Find $f'(x)$ for the following.
   (i) $f(x) = \tan^2 x$.
   (ii) $f(x) = x\sec^2 3x$.
   (iii) $f(x) = x + \cot x - \frac{1}{3}\cot^3 x$.
   (iv) $f(x) = \dfrac{\operatorname{cosec}^2 x}{1 + x^2}$.

2. Determine the following.
   (i) $\dfrac{d}{d\theta}(\sec^3 4\theta)$.
   (ii) $\dfrac{d}{d\theta}\left\{\cot^4\left(\dfrac{\pi}{3} - 3\theta\right)\right\}$.
   (iii) $\dfrac{d}{dx}(x\sec^4 x - \sec^2 x\tan x)$.

3. Sketch the graph of the function $y = \sec x$.

4. Show that the function $y = \sec^2 x$ satisfies the equation

$$\frac{dy}{dx} = 2y\sqrt{y - 1}.$$

Answers

1. (i) $2\tan x\sec^2 x$.

(ii) $\sec^2 3x + 6x \sec^2 3x \tan 3x$.
(iii) $\cot^4 x$.
(iv) $-\dfrac{2 \operatorname{cosec}^2 x \{(1 + x^2) \cot x + x\}}{(1 + x^2)^2}$.

2. (i) $12 \sec^3 4\theta \tan 4\theta$.
(ii) $12 \cot^3 \left(\dfrac{\pi}{3} - 3\theta\right) \operatorname{cosec}^2 \left(\dfrac{\pi}{3} - 3\theta\right)$.
(iii) $4x \sec^4 x \tan x - 2 \sec^2 x \tan^2 x$.

Further Problems for Chapter 2

1. Given that $f(x) = x \sin \pi x$, evaluate $f'(0.25)$.

2. Determine the gradient of the curve

$$y = \frac{\sin x}{x}$$

at the point where $x = 1$. Give your answer correct to three decimal places.

3. The position s of a moving particle at time t is given by

$$s = (1 + t) \cos t - \sin t.$$

Show that the velocity is zero when $t = 0$.

4. Use the quotient rule to derive the result

$$\frac{d}{dx}(\cot x) = -\operatorname{cosec}^2 x.$$

5. Use the function of a function rule to derive the result

$$\frac{d}{dx}(\sec x) = \sec x \tan x.$$

6. Determine

$$\frac{d}{d\theta}(\operatorname{cosec} 4\theta).$$

7. Evaluate

$$\lim_{h \to 0} \left\{ \frac{\sec(x+h) - \sec x}{h} \right\}$$

when $x = \pi/4$.

8. Obtain the equation of the normal to the curve $y = \cot \pi x$ at the point (0.25, 1).

9. Given that $g(z) = (z + \tan z)^2$ eavluate $g'(\pi/4)$.

Answers

1. $\dfrac{1}{4\sqrt{2}}\,(4 + \pi)$.
2. -0.301.
6. $-4 \operatorname{cosec} 4\theta \cot 4\theta$.
7. $\sqrt{2}$.
8. $x - 2\pi y + 2\pi - 0.25 = 0$.
9 $\frac{3}{2}(\pi + 4)$.

3

Higher-order Derivatives

3.1 INTRODUCTION

When the function $f(x)$ is differentiated with respect to x, another function of x, $f'(x)$, is obtained. This function may itself be differentiated with respect to x (assuming that the derivative exists) to form a function of x known as the *second derivative* of $f(x)$ and written $f''(x)$. The notations used for second- and higher-order derivatives are given in section 3.2. Some applications of higher-order derivatives are considered in section 3.3.

3.2 DEFINITIONS AND NOTATION

The **second derivative** of the function $f(x)$ with respect to x is written $f''(x)$ and defined by

$$f''(x) = \frac{d}{dx}\{f'(x)\}.$$

An alternative notation for the second derivative of $y = f(x)$ is

$$\frac{d^2 y}{dx^2} = \frac{d}{dx}\left(\frac{dy}{dx}\right).$$

Similarly the **third derivative** of $y = f(x)$ is

$$f'''(x) = \frac{d}{dx}\{f''(x)\},$$

or, alternatively,

$$\frac{d^3y}{dx^3} = \frac{d}{dx}\left(\frac{d^2y}{dx^2}\right).$$

The *nth derivative* of $y = f(x)$ may be obtained by differentiating $f(x)$ n times and is written $f^{(n)}(x)$ of d^ny/dx^n. (Other notations are in common use, in particular $y', y'', y''', \ldots$ to denote derivatives of $y = f(x)$, and $\dot{y}, \ddot{y}, \dddot{y}, \ldots$ to denote derivatives of $y = f(t)$.)

It is valuable at this stage to appreciate the significance of the *signs* of $f'(x)$ and $f''(x)$ in relation to the graph of the function $y = f(x)$. We have already seen in Chapter 1 that the value of $f'(x)$ at a point on the curve $y = f(x)$ is a measure of the *gradient* of the curve at that point. Fig. 3.1 shows a typical curve with indications of the intervals where the gradient is positive (y *increases* as x is increased) and where the gradient is negative (y *decreases* as x is increased). Also shown are isolated points where the gradient is zero; these are known as **stationary points** and will be considered in detail in Chapter 6.

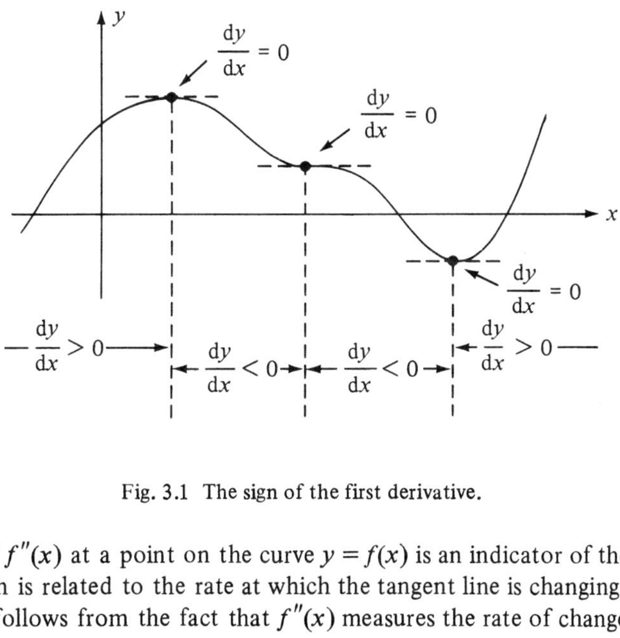

Fig. 3.1 The sign of the first derivative.

The value of $f''(x)$ at a point on the curve $y = f(x)$ is an indicator of the *concavity* of the curve, which is related to the rate at which the tangent line is changing direction as x increases. This follows from the fact that $f''(x)$ measures the rate of change in $f'(x)$ with respect to x, i.e. the rate of change in the gradient of the curve. Fig. 3.2 compares the curves $y = x^2$ and $y = 5x^2$ which have second-derivative values of 2 and 10 respectively (at all points). Note that $y = 5x^2$ is 'more severely curved' than $y = x^2$, in the neighbourhood of the origin of coordinates. This partly reflects the larger value of its second derivative (but see also the calculation of *radius of curvative* in section 3.3).

The *sign* of the second derivative at a point on a curve indicates the *nature* of the curve's concavity at the point. If $f''(x) > 0$ at a point, the curve is said to be **concave up**; if $f''(x) < 0$ at a point, the curve is **concave down** (Fig. 3.3).

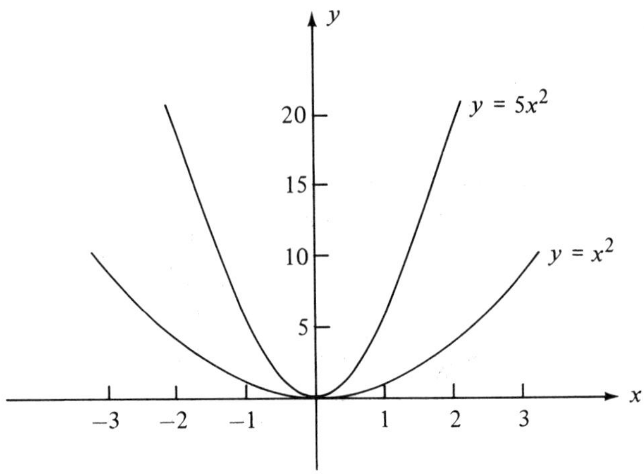

Fig. 3.2

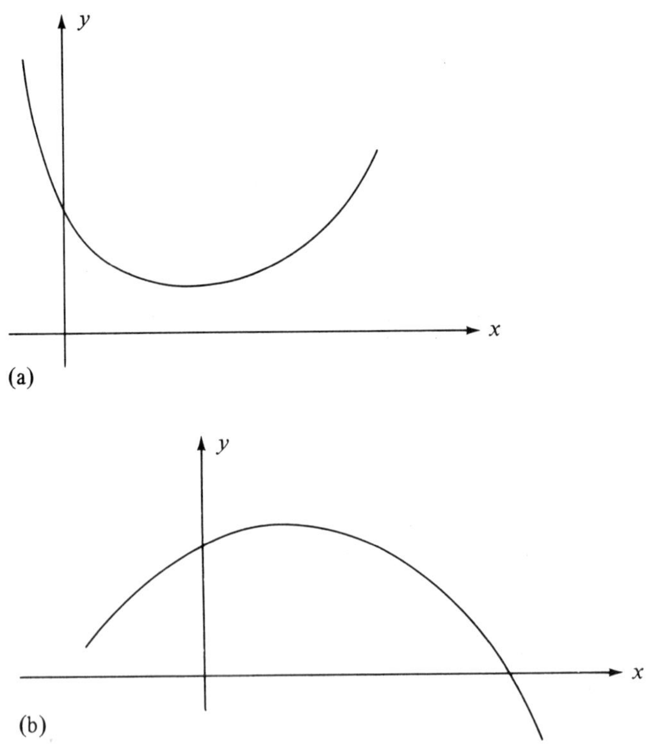

Fig. 3.3 (a) A **concave-up** curve (dy/dx increases as x increases) (b) a **concave-down** curve (dy/dx decreases as x increases).

The statements in the previous paragraph are *not reversible*. If a curve is *concave up* at a point, then $f''(x)$ is *not necessarily positive*; it *may be zero* or *may not exist*, and the only certainty is that $f''(x)$ *cannot be negative*. For example $y = x^4 + x$ is concave up for all x but, at the point (0, 0), $d^2y/dx^2 = 0$. Also the curve $y = x^{4/3}$ is concave up for all x but, at the point (0, 0), the second derivative does not exist (see Example 3.4).

Using the symbol ⇒ to denote 'implies' we can summarize as follows:

$$f''(x) > 0 \Rightarrow y = f(x) \text{ is } \textit{concave up}, \tag{3.1a}$$

$$y = f(x) \text{ is } \textit{concave up} \Rightarrow f''(x) \not< 0, \tag{3.1b}$$

$$f''(x) < 0 \Rightarrow y = f(x) \text{ is } \textit{concave down}, \tag{3.2a}$$

$$y = f(x) \text{ is } \textit{concave down} \Rightarrow f''(x) \not> 0. \tag{3.2b}$$

Fig. 3.4 shows a curve on which the nature of concavity changes many times as x increases. This change in concavity takes place at the isolated points marked A, B, C and D. Such points are called **points of inflexion** and will be considered in detail in Chapter 6. It should be noted that the *gradient* of the curve at a point of inflexion may be *positive* (point D), may be *negative* (point C), may be *zero* (point B) or may *not exist* (point A).

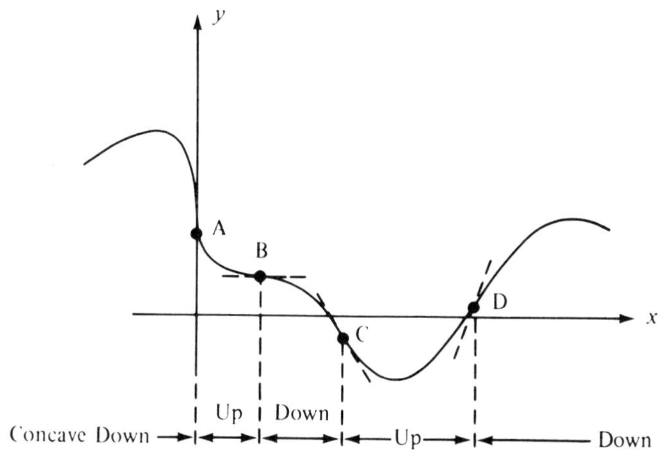

Fig. 3.4 Concavity changes.

Example 3.1

(i) Given that $f(x) = 6x^3 + 4x^2 + 3x + 2$ show that $f''(2) = 80$ and that $f'''(x)$ is independent of x.

(ii) Given

$$y = 4\sqrt{x} - \frac{2}{x},$$

obtain dy/dx, d^2y/dx^2 and d^3y/dx^3.

(iii) For the function $y = \sec^2 x$ show that

$$\frac{d^2y}{dx^2} = 6y^2 - 4y.$$

(i)

$$f(x) = 6x^3 + 4x^2 + 3x + 2.$$

$$f'(x) = 18x^2 + 8x + 3.$$

$$f''(x) = 36x + 8.$$

Therefore,

$$f''(2) = 72 + 8 = 80.$$

$$f'''(x) = 36.$$

Therefore, $f'''(x)$ is independent of x.

(ii)

$$y = 4x^{1/2} - 2x^{-1}.$$

$$\frac{dy}{dx} = 2x^{-1/2} + 2x^{-2}.$$

$$\frac{d^2y}{dx^2} = -x^{-3/2} - 4x^{-3}.$$

$$\frac{d^3y}{dx^3} = \tfrac{3}{2}x^{-5/2} + 12x^{-4}.$$

(iii)

$$y = \sec^2 x = (\sec x)^2.$$

$$\frac{dy}{dx} = 2 \sec x \frac{d}{dx}(\sec x)$$

$$= 2 \sec x \sec x \tan x$$

$$= 2 \sec^2 x \tan x.$$

From the product rule,

$$\frac{d^2y}{dx^2} = 2 \sec^2 x \sec^2 x + 4 \sec^2 x \tan x \tan x$$

$$= 2 \sec^4 x + 4 \sec^2 x \tan^2 x.$$

From equation (0.7),

$$2 \sec^4 x + 4 \sec^2 x \tan^2 x = 2 \sec^4 x + 4 \sec^2 x (\sec^2 x - 1)$$

$$= 6 \sec^4 x - 4 \sec^2 x$$

$$= 6y^2 - 4y,$$

since $y = \sec^2 x$.

Example 3.2
Show that $x = t \sin 2t$ satisfies the equation

$$\frac{d^2x}{dt^2} + 4x = 4 \cos 2t.$$

$x = t \sin 2t$ is differentiated twice using the product rule as follows:

$$\frac{dx}{dt} = t \times 2 \cos 2t + \sin 2t,$$

$$\frac{d^2x}{dt^2} = 2t(-2 \sin 2t) + 2 \cos 2t + 2 \cos 2t$$

$$= -4t \sin 2t + 4 \cos 2t.$$

Now, substitute into the left-hand side of the given equation:

$$-4t \sin 2t + 4 \cos 2t + 4(t \sin 2t) = 4 \cos 2t$$

as required.

Example 3.3
Obtain an expression for the nth derivative of $f(x) = (2x + 3)^k$, where k is a constant. Hence evaluate $f^{(5)}(0)$ when $k = 8$.

$$f'(x) = k(2x + 3)^{k-1}(2).$$

$$f''(x) = k(k-1)(2x + 3)^{k-2}(2^2).$$

$$f'''(x) = k(k-1)(k-2)(2x + 3)^{k-3}(2^3).$$

$$f^{(4)}(x) = k(k-1)(k-2)(k-3)(2x + 3)^{k-4}(2^4).$$

Inspection of the above leads to the expression

$$f^{(n)}(x) = k(k-1)(k-2)\ldots(k-n+1)(2x + 3)^{k-n}(2^n).$$

For the case $f(x) = (2x + 3)^8$,

$$f^{(5)}(0) = (8)(7)(6)(5)(4)(3)^3(2^5)$$

$$= 5\,806\,080.$$

Example 3.4
Sketch the graph of the following.
(i) $y = x^4 + x$.
(ii) $y = x^{4/3}$.

(i) From the scheme set out in section 0.3, we have the following.

(a) **Restrictions.** The domain is all values of x. With respect to the range, clearly, y will have a minimum value since the x^4 term dominates for large x (see (d)).

(b) **Symmetry**. There is no obvious symmetry for this graph.

(c) **Axis intersection.** $y = x(x+1)(x^2 - x + 1)$ and therefore cuts the x axis at $(0, 0)$ and $(-1, 0)$.

(d) **Stationary points.**

$$\frac{dy}{dx} = 4x^3 + 1.$$

$$\frac{d^2y}{dx^2} = 12x^2.$$

From the first derivative we note that the curve has zero *gradient* when $x = -\sqrt[3]{\frac{1}{4}}$. Also $d^2y/dx^2 > 0$ at all points except the *isolated* point $(0, 0)$ where it has the value zero. We conclude that the curve is *always concave up* and that $(-\sqrt[3]{\frac{1}{4}}, -\frac{3}{4}\sqrt[3]{\frac{1}{4}})$ is a *concave-up stationary point*, i.e. a *minimum point.*

(e) **Asymptotes**. There are no asymptotes for this curve (as defined in section 0.3). The sketch is shown in Fig. 3.5(a).

(ii) (a) **Restrictions**. $y = x^{4/3} = (x^{1/3})^4$. Therefore the *domain* is *all* x, and the *range* is $y \geqslant 0$.

(b) **Symmetry**. Since $(-x)^{4/3} = x^{4/3}$, the curve is symmetrical about the y axis.

(c) **Axis intersection.** Only at $(0, 0)$.

(d) **Stationary points.**

$$\frac{dy}{dx} = \frac{4}{3}x^{1/3}.$$

$$\frac{d^2y}{dx^2} = \frac{4}{9}x^{-2/3}.$$

From the first derivative, we note that the curve has zero gradient when $x = 0$, i.e. at the point $(0, 0)$. The range restriction $y \geqslant 0$ implies that the stationary point at $(0, 0)$ is a *minimum point*. Also we note that

$$\frac{d^2y}{dx^2} = \frac{4}{9}(x^{-1/3})^2 > 0$$

at all points except the *isolated* point $(0, 0)$ where it does not exist. (Alternatively, we can say that $d^2y/dx^2 \to \infty$ as $x \to 0$.) It follows that the curve is *concave up* at all points.

(e) **Asymptotes**. There are no asymptotes for this curve (as defined in section 0.3). The sketch is shown in Fig. 3.5(b).

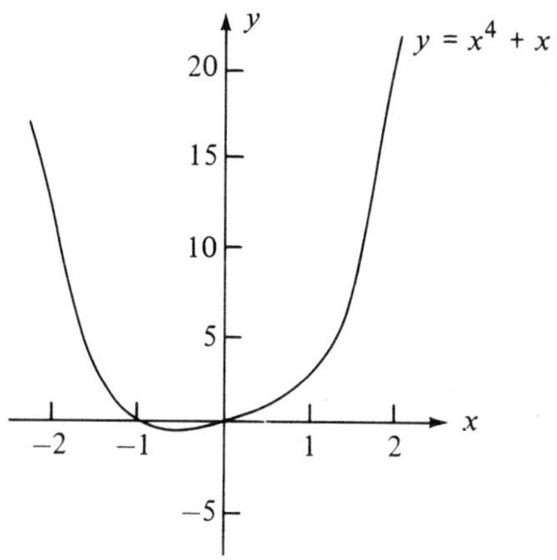

(a)

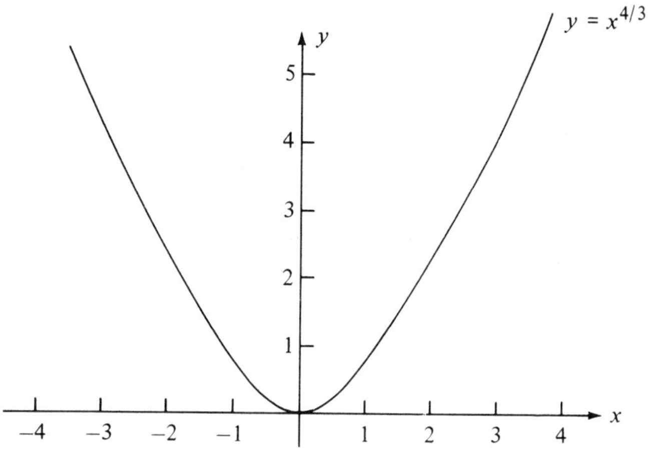

(b)

Fig. 3.5

Problems

1. Find $f'''(x)$ and $f'''(1)$ for the following functions.
   (i) $6x^3 + 4x^2 + 3x + 2$.

   (ii) $\dfrac{1}{x^2} + \dfrac{1}{\sqrt{x}}$.

   (iii) $\sqrt{x + 3}$.

   (iv) $\dfrac{1}{(2x + 1)^2}$.

2. Obtain $d^n y/dx^n$ for the following.
   (i) When $y = x^{-2}$.
   (ii) When $y = (4x + 3)^5$.

3. Find the first and second derivatives of the function

$$y = (x^{-2} - 1)^{-1/2}.$$

4. Show that

$$y = A \sin 2t + B \cos 2t + 2t^2 + t - 1$$

satisfies the equation

$$\frac{d^2y}{dt^2} + 4y = 8t^2 + 4t.$$

5. For the curve

$$y = x^3 + \frac{1}{x^3},$$

determine the following.
   (i) The region where the gradient is positive.
   (ii) The region where the gradient is negative.
   (iii) The region where the curve is concave up.
   (iv) The region where the curve is concave down.
   Sketch the curve.

6. Given that

$$f(x) = x^{3/2} + 4\sqrt{1 - x},$$

determine the domain of the following.
   (i) $f(x)$.
   (ii) $f'(x)$.
   (iii) $f''(x)$.

7. Given that

$$y = t \sin 2t,$$

show that

$$\frac{d^4y}{dt^4} + 8\frac{d^2y}{dt^2} + 16y = 0.$$

Answers

1. (i) 36, 36.

(ii) $-24x^{-1} - (\frac{15}{8})x^{-7/2}, -\frac{207}{8}$.
(iii) $\frac{3}{8}(x+3)^{-5/2}, \frac{3}{256}$.
(iv) $-\dfrac{192}{(2x+1)^5}, -\dfrac{192}{243}$.

2. (i) $(-1)^n (2)(3)(4)\ldots(n+1)(x^{-(n+2)})$.
(ii) $(5)(4)(3)\ldots(6-n)(4x+3)^{5-n}(4^n)$.

3. $\dfrac{dy}{dx} = \dfrac{1}{(1-x^2)^{3/2}}$.

$\dfrac{d^2y}{dx^2} = \dfrac{3x}{(1-x^2)^{5/2}}$.

5. (i) $-\infty < x < -1,\ 1 < x < \infty$.
(ii) $-1 < x < 1,\ x \neq 0$.
(iii) $x > 0$.
(iv) $x < 0$.

6. (i) $0 \leqslant x \leqslant 1$.
(ii) $0 \leqslant x < 1$.
(iii) $0 < x < 1$.

3.3 APPLICATIONS

We have already seen in section 1.3 that, when the position s of a body moving in a straight line is given as a function of time t, then the velocity v is given by ds/dt. Further the acceleration a is given by dv/dt, and it follows that a is the second derivative of s with respect to t, i.e.

$$a = \frac{d^2s}{dt^2}. \tag{3.3}$$

Applications of equation (3.3) arise frequently in dynamics in problems related to Newton's second law, force = mass × acceleration. The solution of such problems often involves the process of *integration* which will be introduced in Chapter 4. However, a few examples based on differentiation will be considered in this section.

Another common application of the second derivative is in the calculation of the **radius of curvature** at a point on a curve. Curvature was mentioned in the previous section and we saw how it was related to the value of the second derivative. The radius ρ of curvature of the curve $y = f(x)$ at a point is calculated from the formula

$$\rho = \left| \frac{\{1 + (dy/dx)^2\}^{3/2}}{d^2y/dx^2} \right|. \tag{3.4}$$

The derivation of equation (3.4) has been given by Thomas and Finney (1979). We simply note that a small value for d^2y/dx^2 implies a large value for ρ, which is consistent with gentle curvature. In problems when dy/dx is small (such as the deflexion of a large beam under loading), we have the approximation

$$\rho = \left| \frac{1}{d^2y/dx^2} \right| . \tag{3.5}$$

Example 3.5

A particle is projected vertically upwards with a velocity of 50 m/s. The height s m reached after t s is given by $s = 50t - 5t^2$. Show that the acceleration is constant. Find the maximum height reached by the particle.

$$s = 50t - 5t^2.$$

$$v = \frac{ds}{dt}$$

$$= 50 - 10t.$$

$$a = \frac{d^2s}{dt^2}$$

$$= -10.$$

Therefore the acceleration is -10 m/s^2 for all t. The particle reaches maximum height when $v = 0$, i.e. when $50 - 10t = 0$. Thus, $t = 5$. Therefore the maximum s is

$$(50t - 5t^2)_{t=5} = 125 \text{ m}.$$

Example 3.6

The position of a moving component is given by

$$s = A \cos nt + B \sin nt,$$

where A, B and n are constants and t is time. Show that the acceleration d^2s/dt^2 is given by $-n^2s$.

$$s = A \cos nt + B \sin nt.$$

$$\frac{ds}{dt} = -An \sin nt + Bn \cos nt.$$

$$\frac{d^2s}{dt^2} = -An^2 \cos nt - Bn^2 \sin nt$$

$$= -n^2 (A \cos nt + B \sin nt)$$

$$= -n^2 s.$$

Example 3.7
Compare the radii of curvature of the curves $y = x^2$, $y = 5x^2$, shown in Fig. 3.2, at the points where $y = 0$ and where $y = 5$.

For $y = x^2$,

$$\frac{dy}{dx} = 2x,$$

$$\frac{d^2y}{dx^2} = 2.$$

and

$$\rho = \left| \frac{(1 + 4x^2)^{3/2}}{2} \right| = \begin{cases} 0.5, & \text{at } (0, 0). \\ 48.12, & \text{at } (\sqrt{5}, 5). \end{cases}$$

For $y = 5x^2$,

$$\frac{dy}{dx} = 10x,$$

$$\frac{d^2y}{dx^2} = 10.$$

and

$$\rho = \left| \frac{(1 + 100x^2)^{3/2}}{10} \right| = \begin{cases} 0.1, & \text{at } (0, 0), \\ 101.5, & \text{at } (1, 5). \end{cases}$$

Example 3.8
Verify that at all points on the semicircle $y = \sqrt{a^2 - x^2}$ the radius of curvature is equal to a.

$$y = (a^2 - x^2)^{1/2}.$$

$$\frac{dy}{dx} = \tfrac{1}{2}(a^2 - x^2)^{-1/2}(-2x).$$

$$= -\frac{x}{\sqrt{a^2 - x^2}}$$

From the quotient rule,

$$\frac{d^2y}{dx^2} = -\frac{\sqrt{a^2 - x^2} \times 1 - x(-x/\sqrt{a^2 - x^2})}{a^2 - x^2}$$

$$= -\frac{a^2}{(a^2 - x^2)^{3/2}}.$$

Therefore, at the point (x, y) on the semicircle,

$$\rho = \left| \frac{\{1 + x^2/(a^2 - x^2)\}^{3/2}}{-a^2/(a^2 - x^2)^{3/2}} \right|$$

$$= \frac{\{a^2/(a^2 - x^2)\}^{3/2}}{a^2/(a^2 - x^2)^{3/2}}$$

$$= \frac{(a^2)^{3/2}}{a^2}$$

$$= a.$$

Problems

1. The distance s m travelled in t s by a body moving in a straight line is given by

$$s = t^3 - t^2.$$

Find the following.
(i) Its velocity after 3 s.
(ii) Its acceleration after 4 s.

2. A stone is thrown vertically upwards with an initial speed of 10 m/s from the top of a dry well which is 20 m deep. Given that the height h of the stone above the top of the well satisfies the equation $h = 10t - 5t^2$, where t is the time measured in seconds, calculate the following.
(i) The time taken for the stone to strike the bottom of the well.
(ii) The velocity of the stone at impact.

3. The angular acceleration of a rotating shaft is given by $\mathrm{d}^2\theta/\mathrm{d}t^2$ where θ measures the rotation of the shaft from some fixed direction. Given that

$$\theta = 0.5 \sin\left(12t + \frac{\pi}{3}\right),$$

find an expression for the angular acceleration in terms of t. What is the angular position when the angular acceleration is maximum?

4. The clockwise angular displacement θ rad of a wheel at time t s is given by the equation $\theta = 2t^3 + 3t^2 + 2$. Find the angular displacement $\Delta\theta$ during the time that the angular velocity $\mathrm{d}\theta/\mathrm{d}t$ changes from 12 to 72 rad/s.

5. Find the radius of curvature of the following curves at the points indicated.
(i) $y = \sqrt{2x}$ at $(2, 2)$.
(ii) $y = \cos x$ at $(0, 1)$.
(iii) $y = \cos x$ at $(\pi/2, 0)$.

6. For what value of y do the curves $y = x^2$ and $y = 5x^2$ have the same radius of curvature at the points (x, y) (see Example 3.7).

7. Fig. 3.6 shows the proposed route for the centre-line of a railway track. The track consists of two straight sections AB and CD joined from B to C by a polynomial curve. It is clearly essential to have no change in gradient and radius of curvature at the points B and C but a third requirement, no change in $\mathrm{d}^3y/\mathrm{d}x^3$, is necessary to ensure a smooth transition from straight track to curved track. Show that the polynomial curve of lowest degree has equation of the form

$$y = (x^2 - 1)(ax^4 + bx^2 + c),$$

and determine the values of a, b and c.

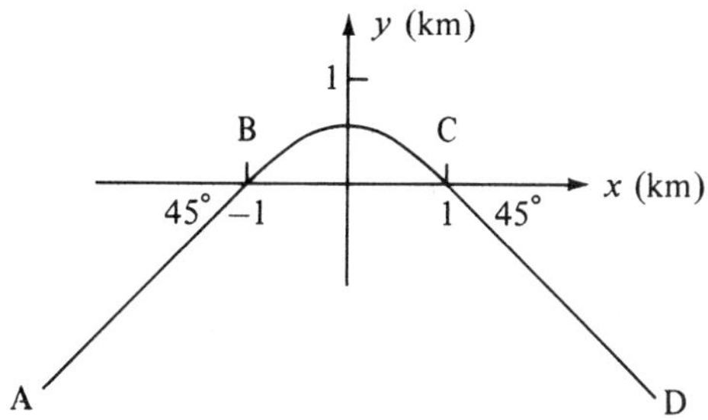

Fig. 3.6

8. A curve $y = f(x)$ is such that $f'(x) = \tan x$ at all points on the curve. Show that the radius of curvature at any point on the curve is given by the value of $|\sec x|$.

Answers

1. (i) 21 m/s.
   (ii) 22 m/s^2.

2. (i) 3.236 s.
   (ii) −22.36 m/s.

3. $\dfrac{\mathrm{d}^2\theta}{\mathrm{d}t^2} = -72 \sin\left(12t + \dfrac{\pi}{3}\right)$.

   Maximum when $\theta = -0.5$.

4. $\Delta\theta = 76$ rad.

5. (i) $5\sqrt{5}$.
   (ii) 1.
   (iii) Infinite. We say that the curvature is 0 (the curvature is defined as $1/\rho$).

6. 0.2317.

7. $a = -\frac{1}{16}$, $b = \frac{1}{4}$, $c = -\frac{11}{16}$.

Further Problems for Chapter 3

1. Evaluate $f''(-1)$ when $f(x) = 12x^{-1}$.

2. Given that $y = t \sin t$ find the value of d^2y/dt^2 when $t = 0$.

3. Given that $y = \cos wx$ verify that

$$\frac{d^n y}{dx^n} = \begin{cases} (-1)^{(n+1)/2}\, w^n \sin wx, & \text{when } n \text{ is odd,} \\ (-1)^{n/2}\, w^n \cos wx, & \text{when } n \text{ is even.} \end{cases}$$

4. Find the interval of values of x for which the curve

$$y = 2x^3 - 12x^2 - 30x + 7$$

is decreasing.

5. Show that the curve

$$y = x^3 + 3x^2 + 15x + 13$$

is increasing for all values of x.

6. Determine the ranges of values of x for which the curve in the previous example is
(i) concave up and
(ii) concave down.
Draw a rough sketch of the curve.

7. The curve $y = f(x)$ has the properties $f(1) = 2$, $f'(1) = 1$, and $f''(1) = -1$. Draw a rough sketch of the curve in the neighbourhood of the point $(1, 2)$.

8. The position s m at time t s of a moving component is given by $s = 0.02 \sin 30t$. Determine the maximum value of the acceleration of the component.

9. Write down the expression for the radius of curvature of the parabola

$$y = ax^2 + bx + c$$

at its stationary point.

10. Give an example to *disprove* the following incorrect statement; 'if $y = f(x)$ is concave down at the point where $x = a$, then $f''(a) < 0$'.

11. Determine the nth derivative with respect to s of the function s^k where k is a constant.

12. Show that the curve

$$y = x^2 + \frac{1}{x^2}$$

is concave up for all non-zero values of x.

Answers

1. -24.
2. 2.
4. $-1 < x < 5$.
6. (i) $x > -1$.
   (ii) $x < -1$.

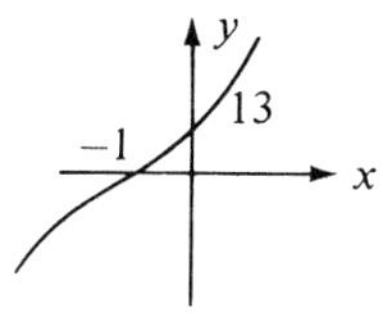

7. 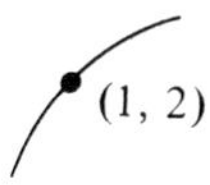

8. 18 m/s^2.
9. $\dfrac{1}{2a}$
10. For example, $y = -x^4$, when $x = 0$.
11. $k(k-1)(k-2)\ldots(k-n+1)s^{k-n}$.

4

Integration

4.1 INTRODUCTION

The process of integration applies to the situation where the derivative $f'(x)$ is given and the function $f(x)$ is required, i.e. integration reverses the process of differentiation. Integration is introduced through the concept of *differentials* and is applied to polynomials and trigonometric functions. A simple method for extending standard integration is given in section 4.3. Integration is applied to certain problems from dynamics in section 4.5.

The applications of integration are considerable and some of these are presented in Chapters 5 and 11.

4.2 DEFINITIONS AND STANDARD FORMS

Fig. 4.1 shows the point P(x, y) on the curve $y = f(x)$, and an arbitrary increment $\mathrm{d}x$. The resulting change in y, the distance QN in the figure, is denoted by Δy such that the coordinates of Q are $(x + \mathrm{d}x, y + \Delta y)$. The tangent to the curve at P is drawn to intersect QN at the point T.

The distance TN = PN $\tan \alpha$ is denoted by the symbol $\mathrm{d}y$. Since PN = $\mathrm{d}x$ and $\tan \alpha = f'(x)$, the gradient of the curve at P, we have $\mathrm{d}y = f'(x)\mathrm{d}x$.

The variable $\mathrm{d}y$ which depends on x *and* $\mathrm{d}x$ plays an important part in the theory of integration and is formally defined below.

Given a function $y = f(x)$ we define a *new variable* $\mathrm{d}y$ as follows:

$$\mathrm{d}y = f'(x)\,\mathrm{d}x.$$

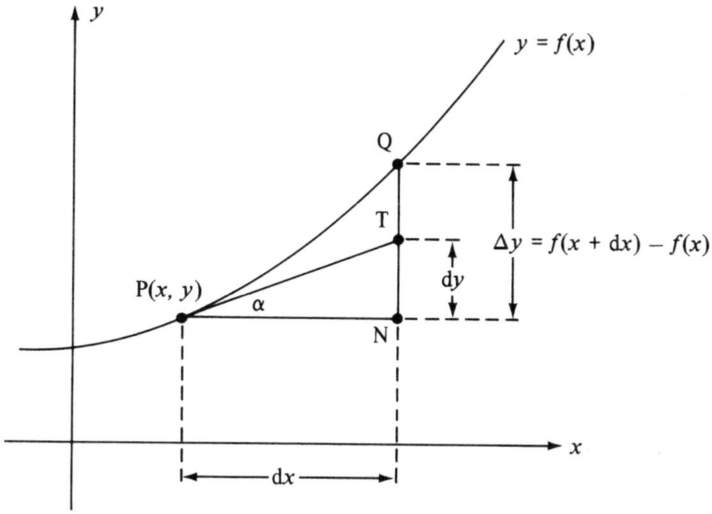

Fig. 4.1

dy is called the **differential** of y. (The variable dx used in the definition is the differential of the particular function $y = x$.)

For example,

(i) if $y = x^3$ then $dy = 3x^2\,dx$, which can be written $d(x^3) = 3x^2\,dx$, and
(ii) if $y = \sin 2x$ then $dy = 2\cos 2x\,dx$, which can be written $d(\sin 2x) = 2\cos 2x\,dx$.

Note that we can divide the differential dy by the differential dx to obtain $dy \div dx = f'(x)$.

Thus the quotient of differentials dy/dx is in fact the same as the *symbol* dy/dx used in Chapter 1.

The definition can be applied to functions of any variable.

For example,

(iii) if $y = t^3 + 2t$, then $dy = (3t^2 + 2)\,dt$, and
(iv) if $u = \sin\theta$ then $du = \cos\theta\,d\theta$.

Differentials have limited application, but the notation is very useful when applying methods of integration.

The **integral** of a given function $f(x)$ is defined as *the function whose differential is* $f(x)\,dx$ and is denoted by

$$\int f(x)\,dx.$$

Thus the symbol $\int$ means 'the function whose differential is'. In other words,

$$\int f(x)\,dx$$

is a function which when differentiated gives $f(x)$, i.e.

$$\int f(x)\,\mathrm{d}x = F(x)$$

implies that

$$F'(x) = f(x).$$

The process of forming the function $\int f(x)\,\mathrm{d}x$ from $f(x)$ is called **integration.** For example,

$$\int 3x^2\,\mathrm{d}x = x^3$$

since

$$\frac{\mathrm{d}}{\mathrm{d}x}(x^3) = 3x^2,$$

or, *in terms of differentials,*

$$\mathrm{d}(x^3) = 3x^2\,\mathrm{d}x.$$

Note that

$$\int 3x^2\,\mathrm{d}x = x^3 + C,$$

where C is an arbitrary constant, since

$$\mathrm{d}(x^3 + C) = 3x^2\,\mathrm{d}x.$$

C is called the **constant of integration** and $x^3 + C$ is referred to as the **indefinite integral** of $3x^2$.

Note also that

$$\mathrm{d}\left(\int 3x^2\,\mathrm{d}x\right) = 3x^2\,\mathrm{d}x$$

and that

$$\int \mathrm{d}(x^3) = x^3 + C,$$

or, in general for *any* suitable function $f(x)$,

$$\mathrm{d}\left\{\int f(x)\,\mathrm{d}x\right\} = f(x)\,\mathrm{d}x, \tag{4.1}$$

$$\int \mathrm{d}\{f(x)\} = f(x) + C. \tag{4.2}$$

Equations (4.1) and (4.2) indicate that, apart from the constant C, the *differential operator* d is the inverse of the *integral operator* $\int$ and vice versa.

The rules for differentiating x^n and the six trigonometric functions derived in Chapters 1 and 2 can now be reversed to establish **standard forms** for integration as follows:

$$\int x^n \, dx = \frac{x^{n+1}}{n+1}, \quad n \neq -1, \tag{4.3}$$

$$\int \cos x \, dx = \sin x, \tag{4.4}$$

$$\int \sin x \, dx = -\cos x, \tag{4.5}$$

$$\int \sec^2 x \, dx = \tan x, \tag{4.6}$$

$$\int \operatorname{cosec} x \cot x \, dx = -\operatorname{cosec} x, \tag{4.7}$$

$$\int \sec x \tan x \, dx = \sec x, \tag{4.8}$$

$$\int \operatorname{cosec}^2 x \, dx = -\cot x. \tag{4.9}$$

Note that

(i) the constant C of integration has been omitted from the above standard results and
(ii) the case $\int x^{-1} \, dx$ will be considered in Chapter 8.

(A list of standard integrals is given in Table 2, at the end of the book.)

The rule for integrating $kf(x)$, where k is a constant, and for integration of the *sum* and *difference* of functions follow from the definition of integration and from equations (1.6) and (1.7), the corresponding rules for differentiation:

$$\int kf(x) \, dx = k \int f(x) \, dx, \tag{4.10}$$

$$\int \{f(x) \pm g(x)\} \, dx = \int f(x) \, dx \pm \int g(x) \, dx. \tag{4.11}$$

Example 4.1

Determine the following.

(i) $\int (5x^3 + 3x + 4) \, dx$.

(ii) $\int (x+1)(x^3 - 2x) \, dx$.

(iii) $\int \left(\sqrt{x} + \frac{3}{\sqrt{x}}\right) dx$.

(iv) $\int \sin 3x \, dx$.

(i) From equation (4.11),

$$\int (5x^3 + 3x + 4) \, dx = \int 5x^3 \, dx + \int 3x \, dx + \int 4 \, dx,$$

from equation (4.10),

$$\int 5x^3\ \mathrm{d}x + \int 3x\ \mathrm{d}x + \int 4\ \mathrm{d}x = 5\int x^3\ \mathrm{d}x + 3\int x\ \mathrm{d}x + 4\int x^0\ \mathrm{d}x,$$

and, from equation (4.3),

$$5\int x^3\ \mathrm{d}x + 3\int x\ \mathrm{d}x + 4\int x^0\ \mathrm{d}x = 5\frac{x^4}{4} + 3\frac{x^2}{2} + 4\frac{x^1}{1} + C$$

$$= \frac{5x^4}{4} + \frac{3x^2}{2} + 4x + C.$$

Clearly, it is not necessary to write down the trivial intermediate steps.

(ii) We have no rule (at this stage) for the integration of the *product* of two functions. Therefore it is necessary to multiply the polynomials as follows:

$$\int (x+1)(x^3 - 2x)\ \mathrm{d}x = \int (x^4 + x^3 - 2x^2 - 2x)\ \mathrm{d}x$$

$$= \frac{x^5}{5} + \frac{x^4}{4} - \frac{2x^3}{3} - x^2 + C.$$

(iii)

$$\int \left(\sqrt{x} + \frac{3}{\sqrt{x}}\right) \mathrm{d}x = \int (x^{1/2} + 3x^{-1/2})\ \mathrm{d}x$$

$$= \frac{x^{3/2}}{3/2} + 3\frac{x^{1/2}}{1/2} + C$$

$$= \tfrac{2}{3}x^{3/2} + 6x^{1/2} + C.$$

(iv) Integrals such as $\int \sin 3x\ \mathrm{d}x$ may be determined using the method of substitution described in section 4.3. However, in this case the simplest method is to *guess* the answer and to check that the derivative of your guess is $\sin 3x$. If not, a suitable (constant) factor may be required. Consideration of the standard form (4.5) suggests that the integral may be $-\cos 3x$. Now,

$$\frac{\mathrm{d}}{\mathrm{d}x}(-\cos 3x) = 3\sin 3x$$

and

$$\frac{\mathrm{d}}{\mathrm{d}x}(-\tfrac{1}{3}\cos 3x) = \sin 3x.$$

It follows that

$$\int \sin 3x\ \mathrm{d}x = -\tfrac{1}{3}\cos 3x + C.$$

Example 4.2
Obtain the following functions.

(i) $\int \sec\theta \tan\theta \, d\theta.$

(ii) $\int (t^2 + 1)^2 \, dt.$

(iii) $\int \dfrac{(y-2)^2}{y^4} \, dy.$

(i) $\int \sec\theta \tan\theta \, d\theta = \sec\theta + C$

since

$$\frac{d}{d\theta}(\sec\theta) = \sec\theta \tan\theta.$$

This is simply the standard form (4.8).

(ii)
$$\int (t^2 + 1)^2 \, dt = \int (t^4 + 2t^2 + 1) \, dt$$
$$= \frac{t^5}{5} + \frac{2t^3}{3} + t + C.$$

(iii)
$$\int \frac{(y-2)^2}{y^4} \, dy = \int \frac{y^2 - 4y + 4}{y^4} \, dy$$
$$= \int (y^{-2} - 4y^{-3} + 4y^{-4}) \, dy$$
$$= -y^{-1} + 2y^{-2} - \tfrac{4}{3} y^{-3} + C.$$

Example 4.3
Show that

$$\int (\tan^3 x + \tan x) \, dx = \tfrac{1}{2} \tan^2 x + C.$$

In this case the result of the integration is *given* and all we need to do is to *check* that

$$\frac{d}{dx}(\tfrac{1}{2} \tan^2 x) = \tan^3 x + \tan x.$$

Now

$$\frac{d}{dx}(\tfrac{1}{2} \tan^2 x) = \frac{d}{dx}\{\tfrac{1}{2} (\tan x)^2\}$$
$$= \tan x \sec^2 x.$$

From equation (0.7),

$$\tan x \sec^2 x = \tan x\,(1 + \tan^2 x),$$
$$= \tan x + \tan^3 x$$

as required.

Problems

1. Find the differentials of the following functions.
   (i) $x^3 \sin x$.
   (ii) $\sin^2 \theta$.
   (iii) $(t + 1)^4$.
   (iv) $(y^2 - y)^3$.

2. Determine the following integrals.

   (i) $\int (x^3 - 3x^2 + 2x - 7)\, dx$.

   (ii) $\int \left\{ \frac{(\sqrt{x} + x^{-1})^2}{\sqrt{x}} - \frac{2}{x} \right\} dx$.

   (iii) $\int \cos 7x\, dx$.

   (iv) $\int 2 \sin \left(2\theta + \frac{\pi}{3} \right) d\theta$.

   (v) $\int \sec^2 (1 - x)\, dx$.

   (vi) $\int (2y + 1)^3\, dy$.

   (vii) $\int (y^2 + 1)^3\, dy$.

3. Prove the following.

   (i) $\int (x^2 + 1)^5\, x\, dx = \frac{1}{12} (x^2 + 1)^6 + C$.

   (ii) $\int x \sin x\, dx = -x \cos x + \sin x + C$.

Answers

1. (i) $(3x^2 \sin x + x^3 \cos x)\, dx$.
   (ii) $2 \sin \theta \cos \theta\, d\theta$.
   (iii) $4(t + 1)^3\, dt$.
   (iv) $3(y^2 - y)^2\, (2y - 1)\, dy$.

2. (i) $\dfrac{x^4}{4} - x^3 + x^2 - 7x + C.$

(ii) $\frac{2}{3}x^{3/2} - \frac{2}{3}x^{-3/2} + C.$

(iii) $\frac{1}{7}\sin 7x + C.$

(iv) $-\cos\left(2\theta + \dfrac{\pi}{3}\right) + C.$

(v) $-\tan(1 - x) + C.$

(vi) $2y^4 + 4y^3 + 3y^2 + y + C$ or $\frac{1}{8}(2y + 1)^4 + C.$

(vii) $\dfrac{y^7}{7} + \dfrac{3y^5}{5} + y^3 + y + C.$

4.3 SIMPLE SUBSTITUTIONS

Many integrals can be reduced to one of the standard forms by a *change in variable.* For example, consider again $\int \sin 3x \, dx$ (Example 4.1 (iv)). Substitute $u = 3x$. Therefore $du = 3\, dx$ and $dx = \frac{1}{3}\, du$. Thus

$$\int \sin 3x \, dx = \int \sin u \, \tfrac{1}{3}\, du$$

From equation (4.10),

$$\int \sin u \, \tfrac{1}{3}\, du = \tfrac{1}{3}\int \sin u \, du$$

which in the standard form (4.5) is

$$-\tfrac{1}{3}\cos u + C = -\tfrac{1}{3}\cos 3x + C.$$

Note that, when making the change in variable, we replace a differential $f(x)\,dx$ by another differential $g(u)\,du$. Thus, when making a substitution, we must remember to replace both the $f(x)$ *and* the dx parts with expressions involving the new variable. The technique is known as **integration by substitution** and will be given further consideration in Chapter 10.

Example 4.4

Use substitutions to determine the following integrals (all appeared in section 4.2).

(i) $\int \cos 7x \, dx.$

(ii) $\int \sin\left(2\theta + \dfrac{\pi}{3}\right) d\theta.$

(iii) $\int \sec^2 (1 - x)\, dx.$

(iv) $\int (2y+1)^3\,dy$.

(v) $\int (x^2+1)^5 x\,dx$.

(i) Put $u = 7x$; $du = 7\,dx$, i.e. $dx = \frac{1}{7}\,du$. Therefore,

$$\begin{aligned}\int \cos 7x\,dx &= \int \cos u\,\tfrac{1}{7}\,du \\ &= \tfrac{1}{7}\int \cos u\,du \\ &= \tfrac{1}{7}\sin u + C \\ &= \tfrac{1}{7}\sin 7x + C.\end{aligned}$$

(ii) Put $u = 2\theta + \pi/3$; $du = 2\,d\theta$, i.e. $d\theta = \frac{1}{2}\,du$, Therefore,

$$\begin{aligned}\int \sin\left(2\theta + \frac{\pi}{3}\right) dx &= \tfrac{1}{2}\int \sin u\,du \\ &= -\tfrac{1}{2}\cos u + C \\ &= -\tfrac{1}{2}\cos\left(2\theta + \frac{\pi}{3}\right) + C.\end{aligned}$$

(iii) Put $u = 1 - x$; $du = -\,dx$, i.e. $dx = -\,du$. Therefore,

$$\begin{aligned}\int \sec^2(1-x)\,dx &= -\int \sec^2 u\,du \\ &= -\tan u + C \\ &= -\tan(1-x) + C.\end{aligned}$$

(iv) Put $u = 2y + 1$; $du = 2\,dy$, i.e. $dy = \frac{1}{2}\,du$. Therefore,

$$\begin{aligned}\int (2y+1)^3\,dy &= \tfrac{1}{2}\int u^3\,du \\ &= \tfrac{1}{2}\,\frac{u^4}{4} + C \\ &= \tfrac{1}{8}(2y+1)^4 + C.\end{aligned}$$

Note that it is possible to determine the integral $\int (2y+1)^3\,dy$ by the 'guesswork' method described in Example 4.1 (iv). However, it was not possible to guess the answer to $\int (y^2+1)^3\,dy$, problem 2 (vii) in section 4.2. Further, the substitution $u = y^2 + 1$, $du = 2y\,dy$, does not help because there is no simple expression for dy in terms of du and u. To determine $\int (y^2+1)^3\,dy$, it is necessary to expand the binomial $(y^2+1)^3$ and integrate term by term. In general the 'guesswork' method should only be used when integrating functions of a *linear* expression such as $2y + 1, 1 - x, 3x, 2\theta + \pi/3$.

(v) Put $u = x^2 + 1$; $du = 2x\,dx$, i.e. $x\,dx = \frac{1}{2}\,du$. Therefore,

$$\begin{aligned}\int (x^2+1)^5 x\,dx &= \tfrac{1}{2}\int u^5\,du\\ &= \tfrac{1}{2}\,\frac{u^6}{6} + C\\ &= \tfrac{1}{12}(x^2+1)^6 + C.\end{aligned}$$

(Note that the presence of the multiplier x in the function to be integrated allows the method of substitution to succeed. This example is a simple case of the general type

$$\int F\{f(x)\}\,f'(x)\,dx = \int F(u)\,du.$$

Further examples of this type follow in Examples 4.5 and 4.6.)

Example 4.5

Carry out the following integrations.

(i) $\displaystyle\int \frac{x^2}{(2+x^3)^2}\,dx.$

(ii) $\displaystyle\int x \sin\left(x^2 - \frac{\pi}{4}\right) dx.$

(iii) $\displaystyle\int (1 + \cos^2 u) \sin u\,du.$

(i) We note that the multiplier x^2 is proportional to the derivative of the function $2 + x^3$. Therefore, put $u = 2 + x^3$; $du = 3x^2\,dx$, i.e. $x^2\,dx = \frac{1}{3}\,du$.

$$\begin{aligned}\int \frac{x^2}{(2+x^3)^2}\,dx &= \tfrac{1}{3}\int \frac{du}{u^2}\\ &= \tfrac{1}{3}\int u^{-2}\,du\\ &= -\tfrac{1}{3}u^{-1} + C\\ &= -\frac{1}{3(2+x^3)} + C.\end{aligned}$$

(ii) We note that the multiplier x is proportional to the derivative of the function $x^2 - \pi/4$. Therefore, put $u = x^2 - \pi/4$; $du = 2x\,dx$, i.e. $x\,dx = \frac{1}{2}\,du$.

$$\int x \sin\left(x^2 - \frac{\pi}{4}\right) \mathrm{d}x = \tfrac{1}{2}\int \sin u \, \mathrm{d}u$$

$$= -\tfrac{1}{2}\cos u + C$$

$$= -\tfrac{1}{2}\cos\left(x^2 - \frac{\pi}{4}\right) + C.$$

(iii) Since the multiplier $\sin u$ is proportional to the derivative of the function $\cos u$, we choose the substitution $v = \cos u$; then $\mathrm{d}v = -\sin u \, \mathrm{d}u$, i.e. $\sin u \, \mathrm{d}u = -\mathrm{d}v$.

$$\int (1 + \cos^2 u) \sin u \, \mathrm{d}u = -\int (1 + v^2)\, \mathrm{d}v$$

$$= -(v + \tfrac{1}{3}v^3) + C$$

$$= -\cos u - \tfrac{1}{3}\cos^3 u + C.$$

Example 4.6

Use a substitution to show that

$$\int \{f(x)\}^n f'(x)\, \mathrm{d}x = \frac{\{f(x)\}^{n+1}}{n+1} + C, \quad n \neq -1.$$

Put $u = f(x)$; $\mathrm{d}u = f'(x)\,\mathrm{d}x$. Therefore,

$$\int \{f(x)\}^n f'(x)\, \mathrm{d}x = \int u^n \, \mathrm{d}u$$

$$= \frac{u^{n+1}}{n+1} + C, \quad n \neq -1$$

$$= \frac{\{f(x)\}^{n+1}}{n+1} + C.$$

Example 4.7

Use the substitution $u = x + 1$ to determine the function

$$\int \frac{x}{(x+1)^4}\, \mathrm{d}x.$$

The substitution given reduces the denominator of

$$\frac{x}{(x+1)^4}$$

to a single term so that division can be carried out prior to integration.

$$u = x + 1, \qquad \mathrm{d}u = \mathrm{d}x, \qquad x = u - 1.$$

Therefore,

$$\int \frac{x}{(x+1)^4}\,dx = \int \frac{u-1}{u^4}\,du$$

$$= \int (u^{-3} - u^{-4})\,du$$

$$= -\tfrac{1}{2}u^{-2} + \tfrac{1}{3}u^{-3} + C$$

$$= -\tfrac{1}{2}(x+1)^{-2} + \tfrac{1}{3}(x+1)^{-3} + C.$$

Problems

1. Determine the following integrals.

(i) $\int (2x+3)^8\,dx.$

(ii) $\int (x^3+4)^2 x^2\,dx.$

(iii) $\int (x^3+4)^2\,dx.$

(iv) $\int \cos\left(\frac{\pi}{2} - 7x\right)dx.$

(v) $\int \sin^2 x \cos x\,dx.$

(vi) $\int x^2\sqrt{x^3+1}\,dx.$

(vii) $\int \frac{\sin\sqrt{x}}{\sqrt{x}}\,dx.$

(viii) $\int 3\sqrt{27-y}\,dy.$

(ix) $\int \frac{3t^2}{\sqrt{t^3+4}}\,dt.$

(x) $\int \frac{7x+3}{(x+1)^4}\,dx.$

(xi) $\int \frac{\sin x}{\sqrt{\cos^3 x}}\,dx.$

(xii) $\int \frac{2u+3}{\sqrt{u^2+3u+41}}\,du.$

2. Given that

$$\int f(x)\,\mathrm{d}x = F(x) + C,$$

prove the following.

(i) $\int f(ax+b)\,\mathrm{d}x = \frac{1}{a}\,F(ax+b) + C.$

(ii) $\int f\{g(x)\}\,g'(x)\,\mathrm{d}x = F\{g(x)\} + C.$

Answers

(i) $\frac{1}{18}(2x+3)^9 + C.$

(ii) $\frac{1}{9}(x^3+4)^3 + C.$

(iii) $\frac{1}{7}x^7 + 2x^4 + 16x + C.$

(iv) $-\frac{1}{7}\sin\left(\frac{\pi}{2} - 7x\right) + C.$

(v) $\frac{1}{3}\sin^3 x + C.$

(vi) $\frac{2}{9}(x^3+1)^{3/2} + C.$

(vii) $-2\cos(\sqrt{x}) + C.$

(viii) $-2(27-y)^{3/2} + C.$

(ix) $2\sqrt{t^3+4} + C.$

(x) $\frac{4}{3}(x+1)^{-3} - \frac{7}{2}(x+1)^{-2} + C.$

(xi) $\frac{2}{\sqrt{\cos x}} + C.$

(xii) $2\sqrt{u^2+3u+41} + C.$

4.4 USE OF TRIGONOMETRIC IDENTITIES

The integration of trigonometric expressions is often simplified by the application of identities such as those set out in section 0.7.

Example 4.8

1 Determine the following integrals.

(i) $\int \cos^2 x\,\mathrm{d}x.$

(ii) $\int \sin^2 2x\,\mathrm{d}x.$

(iii) $\int \sin x \sin 3x\,\mathrm{d}x.$

(iv) $\int \cos^3 x \, dx$.

(v) $\int \cos^3 x \sin x \, dx$.

(i) $\int \cos^2 x \, dx = \int \frac{1}{2}(1 + \cos 2x) \, dx$

$$= \tfrac{1}{2}(x + \tfrac{1}{2}\sin 2x) + C$$
$$= \tfrac{1}{2}x + \tfrac{1}{4}\sin 2x + C.$$

(ii) $\int \sin^2 2x \, dx = \int \frac{1}{2}(1 - \cos 4x) \, dx$

$$= \tfrac{1}{2}(x - \tfrac{1}{4}\sin 4x) + C$$
$$= \tfrac{1}{2}x - \tfrac{1}{8}\sin 4x + C.$$

(iii) $\int \sin x \sin 3x \, dx = \int \frac{1}{2}(\cos 2x - \cos 4x) \, dx$

$$= \tfrac{1}{2}(\tfrac{1}{2}\sin 2x - \tfrac{1}{4}\sin 4x) + C$$
$$= \tfrac{1}{4}\sin 2x - \tfrac{1}{8}\sin 4x + C.$$

(iv) $\int \cos^3 x \, dx = \int \cos^2 x \cos x \, dx$

$$= \int (1 - \sin^2 x) \cos x \, dx.$$

Now put $u = \sin x$; $du = \cos x \, dx$. Therefore,

$$\int \cos^3 x \, dx = \int (1 - u^2) \, du$$
$$= u - \frac{u^3}{3} + C$$
$$= \sin x - \tfrac{1}{3}\sin^3 x + C.$$

Note that, in general, when integrating an *even* power of $\cos x$ (or $\sin x$), we require the identity expressing $\cos^2 x$ (or $\sin^2 x$) in terms of $\cos 2x$ but, when integrating an *odd* power of $\cos x$ (or $\sin x$), we proceed as above.

(v) $\int \cos^3 x \sin x \, dx$ can be obtained by a technique similar to that used in (iv), but it is simpler to substitute $u = \cos x$ and $du = -\sin x \, dx$ to give

$$\int \cos^3 x \sin x \, dx = -\int u^3 \, du$$
$$= -\frac{u^4}{4} + C$$
$$= -\tfrac{1}{4}\cos^4 x + C.$$

Problems

1. Carry out the following integrations.

(i) $\int \sin x \cos x \, dx.$

(ii) $\int (\sin x \cos x)^2 \, dx.$

(iii) $\int \cos^2 (1 - 3x) \, dx.$

(iv) $\int \cos 4x \cos 3x \, dx.$

(v) $\int \cos 4x \sin 3x \, dx.$

(vi) $\int \tan^2 x \, dx.$

(vii) $\int (2 + \cos x)^2 \, dx.$

(viii) $\int \cos^4 x \, dx.$

(ix) $\int \sin^5 2x \, dx.$

(x) $\int x \sin^2 (x^2) \, dx.$

Answers

1. (i) $-\frac{1}{4} \cos 2x + C.$

(ii) $\dfrac{x}{8} - \frac{1}{32} \sin 4x + C.$

(iii) $\dfrac{x}{2} - \frac{1}{12} \sin (2 - 6x) + C.$

(iv) $\frac{1}{14} \sin 7x + \frac{1}{2} \sin x + C.$

(v) $-\frac{1}{14} \cos 7x + \frac{1}{2} \cos x + C.$

(vi) $\tan x - x + C.$

(vii) $\dfrac{9x}{2} + 4 \sin x + \frac{1}{4} \sin 2x + C.$

(viii) $\dfrac{3x}{8} + \frac{1}{4} \sin 2x + \frac{1}{32} \sin 4x + C.$

(ix) $-\frac{1}{2} \cos 2x + \frac{1}{3} \cos^3 2x - \frac{1}{10} \cos^5 2x + C.$

(x) $\dfrac{x^2}{4} - \frac{1}{8} \sin (2x^2) + C.$

4.5 APPLICATION TO PROBLEMS IN DYNAMICS

In section 1.3 we saw that, for motion in a straight line, the distance travelled s, the velocity v and the acceleration a are related by $v = \mathrm{d}s/\mathrm{d}t$, $a = \mathrm{d}v/\mathrm{d}t = \mathrm{d}^2s/\mathrm{d}t^2$ where t is time. It follows that, provided that a is known as a function of t, v may be obtained by the process of integration from the differential $\mathrm{d}v = (\mathrm{d}v/\mathrm{d}t)\,\mathrm{d}t = a\,\mathrm{d}t$, i.e.

$$v = \int a\,\mathrm{d}t. \tag{4.12}$$

Similarly,

$$s = \int v\,\mathrm{d}t. \tag{4.13}$$

Equations (4.12) and (4.13) express v and s as *indefinite* integrals, but there is nothing indefinite about a real problem in dynamics. Real problems always include sufficient additional information so that the *constant of integration* may be evaluated for each integration carried out. For example, if the acceleration is given and the distance travelled is required, then *two integrations* must be carried out; therefore, *two additional conditions* must be supplied in the statement of the problem. These conditions usually take the form of **initial conditions** or **boundary conditions.** Initial conditions supply the values of distance and velocity at the start of motion (usually $t = 0$). Boundary conditions (which may include initial conditions) supply information at significant stages in the motion.

Example 4.9

A stone is dropped from the highest point of a river bridge and hits the river surface 2.5 s later. Neglecting air resistance and given that the acceleration due to gravity is 9.81 m/s^2, calculate the height of the bridge.

The equation of motion for this *constant-acceleration* situation is

$$\frac{\mathrm{d}v}{\mathrm{d}t} = 9.81$$

Therefore,

$$\begin{aligned} v &= \int 9.81\,\mathrm{d}t \\ &= 9.81t + C_1, \end{aligned}$$

i.e.

$$\frac{\mathrm{d}s}{\mathrm{d}t} = 9.81t + C_1.$$

Integrating again gives

$$\begin{aligned} s &= \int (9.81t + C_1)\,\mathrm{d}t \\ &= 4.905t^2 + C_1 t + C_2. \end{aligned}$$

Normally C_1 and C_2 are *arbitrary* constants, but in this real problem they must take particular values to satisfy the given initial conditions. Suppose that we measure distance, positive downwards, from the top of the bridge such that $s = 0$ at the start of the motion, i.e. $s = 0$ when $t = 0$ (the first initial conditon). Also we are told that the stone was dropped from rest, i.e. $v = 0$ when $t = 0$ (the second initial conditon).

To satisfy $s = 0$ when $t = 0$, we must have $C_2 = 0$. To satisfy $v = 0$ when $t = 0$ we must have $C_1 = 0$. Therefore, $s = 4.905t^2$.

Finally, if h is the height of the bridge, then we are given that $s = h$ when $t = 2.5$.

It follows that

$$\begin{aligned} h &= 4.905 \times 2.5^2 \\ &= 30.66 \text{ m}. \end{aligned}$$

Example 4.10

A test vehicle for a rocket engine runs on a straight horizontal track. The vehicle starts from rest at time $t = 0$ s and the engine thrust T N during the time interval $0 \leqslant t \leqslant 2$ is given by

$$\frac{T}{1000} = \begin{cases} t, & \text{when } 0 \leqslant t \leqslant 1, \\ 1, & \text{when } 1 \leqslant t \leqslant 2. \end{cases}$$

Given that the mass m of the vehicle plus engine is 500 kg and that Newton's second law $T = m\,(\mathrm{d}^2s/\mathrm{d}t^2)$ applies, calculate the position and velocity of the vehicle at the following times.

(i) When $t = 1$.
(ii) When $t = 2$.
(Neglect the friction on the track.)

(i) In the interval $0 \leqslant t \leqslant 1$

$$\frac{\mathrm{d}^2s}{\mathrm{d}t^2} = \frac{T}{m} = \frac{1000t}{500} = 2t.$$

Integrating once,

$$\frac{\mathrm{d}s}{\mathrm{d}t} = t^2 + C_1$$

and $C_1 = 0$ since the velocity $v = 0$ when $t = 0$. Integrating again,

$$s = \frac{t^3}{3} + C_2$$

and $C_2 = 0$ since $s = 0$ when $t = 0$, i.e. $s = t^3/3$ and $v = \mathrm{d}s/\mathrm{d}t = t^2$.

Therefore, when $t = 1$,

$$s = \tfrac{1}{3}\, m$$

and

$$v = 1 \text{ m/s}.$$

(ii) In the interval $1 \leqslant t \leqslant 2$,

$$\frac{d^2s}{dt^2} = \frac{1000}{500} = 2.$$

and, from (i) two conditions have to be satisfied; $s = \frac{1}{3}$ when $t = 1$ and $ds/dt = 1$ when $t = 1$. Integrating once,

$$\frac{ds}{dt} = 2t + C_3$$

and $C_3 = -1$ to satisfy the condition at $t = 1$. Integrating again,

$$s = t^2 - t + C_4$$

and $C_4 = \frac{1}{3}$ to satisfy the condition at $t = 1$, i.e. $s = t^2 - t + \frac{1}{3}$ and $v = ds/dt = 2t - 1$. Therefore, when $t = 2$,

$$s = \tfrac{7}{3} \text{ m}$$

and

$$v = 3 \text{ m/s}.$$

Example 4.11

A particle, having an initial velocity u m/s, travels in a straight line with a constant acceleration a m/s^2. Derive expressions for its velocity v m/s and displacement s m from its initial position in terms of elapsed time t s.

The equation of motion is

$$\frac{d^2s}{dt^2} = a.$$

Therefore,

$$\frac{ds}{dt} = at + C_1$$

and

$$s = \tfrac{1}{2}at^2 + C_1 t + C_2.$$

The initial conditions are $s = 0$ when $t = 0$ and $ds/dt = u$ when $t = 0$. It follows that $C_1 = u$ and $C_2 = 0$. Therefore the velocity $v = u + at$, and the displacement $s = ut + \frac{1}{2}at^2$.

The above formulae for v and s are extremely well known but have limited application. They only apply for *constant-acceleration* situations such as Example 4.9.

Problems

1. A stone is thrown vertically downwards from the highest point of a bridge. It hits the river surface 50 m below 2 s later. Calculate the initial velocity of the stone. (Neglect air resistance and take the acceleration due to gravity to be 9.81 m/s^2.)

2. A rocket is launched from rest at sea level with an exhaust flow rate which gives it a constant vertical acceleration of magnitude g. A second rocket is launched 10 s later with a constant vertical acceleration of $2g$. Determine the altitude at which the second rocket overtakes the first. (Assume that $g = 9.81$ m/s^2.)

3. If in Example 4.10 the thrust is modified such that

$$\frac{T}{1000} = \begin{cases} kt, & \text{when } 0 \leqslant t \leqslant 1, \\ 1, & \text{when } 1 < t \leqslant 2, \end{cases}$$

determine the value of the constant k which gives a velocity of 5 m/s when $t = 2$. Calculate the distance travelled by the vehicle when $t = 2$ for this value of k.

4. A body of mass m is acted upon by a force $F \cos nt$ where F and n are constants. It starts from rest at $t = 0$. If the displacement of the body after a time t is x, prove that

$$x = \frac{F}{n^2 m}(1 - \cos nt).$$

(Assume that force = mass × acceleration.)

5. The angular velocity of a gear is given by

$$\frac{d\theta}{dt} = 6 - 3t^2 \text{ rad/s},$$

measured positive clockwise. Find the angular displacement θ when $t = 3$ s relative to the position of the gear when $t = 0$.

Answers

1. 15.19 m/s.
2. 5.72 km.
3. $k = 3$, 5 m.
5. -9 rad (or $-(9 - 2\pi)$ rad).

Further problems for Chapter 4

1. Write down the differential of

$$y = x^4 + 3x^2 - 2.$$

2. Given that $d(2 \cos 3t) = f(t)\, dt$ determine $f(t)$.

3. Without attempting to carry out the integration, determine

$$\frac{d}{dx}\left(\int \tan^2 x \, dx\right).$$

4. Determine the function

$$\int (2 \sin x + \sec^2 x)\, dx.$$

5. Given that $f'(t) = \sqrt{t} + \cos 3t - 4$, determine $f(t)$.

6. Obtain the indefinite integral of the function $(2x^2 + 1)^2$.

7. Prove that

$$\int \frac{4(x+1)}{(x^2+2x)^3}\, dx = -\frac{1}{(x^2+2x)^2} + C.$$

8. Determine the function $\int \sin 7\theta \, d\theta$.

9. Obtain the indefinite integral of the function $(2x + 1)^4$.

10. Carry out the integration

$$\int y\sqrt{1+y^2}\, dy.$$

11. Determine the function

$$\int \sin^2\left(\frac{x}{2}\right) dx.$$

12. Given that

$$df = \sin^2\left(\frac{u}{2}\right) \cos\left(\frac{u}{2}\right) du$$

determine the function $f(u)$.

13. The velocity of a component is given by

$$\frac{ds}{dt} = 3 \sin t$$

and the condition $s = 1$ when $t = 0$ must be satisfied. Express s in terms of t.

14. The acceleration of a moving vehicle is 2 $\mathrm{m/s^2}$. Given that the initial values of velocity v and position s are 0 m/s and 10 m respectively, express s in terms of t.

Answers

1. $(4x^3 + 6x)\,\mathrm{d}x$.
2. $-6 \sin 3t$.
3. $\tan^2 x$.
4. $-2 \cos x + \tan x + C$.
5. $\frac{2}{3} t^{3/2} + \frac{1}{3} \sin 3t - 4t + C$.
6. $\dfrac{4x^5}{5} + \dfrac{4x^3}{3} + x + C$.
8. $-\frac{1}{7} \cos 7\theta + C$.
9. $\frac{1}{10} (2x + 1)^5 + C$.
10. $\frac{1}{3} (1 + y^2)^{3/2} + C$.
11. $\frac{1}{2} (x - \sin x) + C$.
12. $\frac{2}{3} \sin^3 \left(\dfrac{u}{2}\right) + C$.
13. $4 - 3 \cos t$.
14. $t^2 + 10$.

5

Definite Integrals

5.1 INTRODUCTION

A relationship is derived between the indefinite integral $\int f(x)\,\mathrm{d}x$ and the area lying between the curve $y = f(x)$ and the x-axis. This relationship leads to the definition of the *definite integral* of a function. In section 5.3, it is shown that the definite integral is the limit of a summation, a fundamental result used in many applications of integration. In this chapter, integration is applied to the calculation of areas and volumes. Further applications are considered in Chapter 11.

5.2 THE AREA UNDER A CURVE

Suppose the function $y = f(x)$ is defined for $x \geqslant a$ and is such that $f(x) \geqslant 0$.

Consider the function $A(x)$ which is equal to the shaded area in Fig. 5.1. We refer to this area as the **area under the curve** $y = f(x)$ between the values a and x of the independent variable. The unshaded column in Fig. 5.1 is the area under the curve between the values x and $x + \Delta x$.

We denote this area by ΔA and, clearly,

$$\Delta A = A(x + \Delta x) - A(x).$$

When Δx is small, the area ΔA is approximately equal to $f(x)\,\Delta x$, (the height of the column multiplied by the width), i.e.

$$f(x) \approx \frac{A(x + \Delta x) - A(x)}{\Delta x}.$$

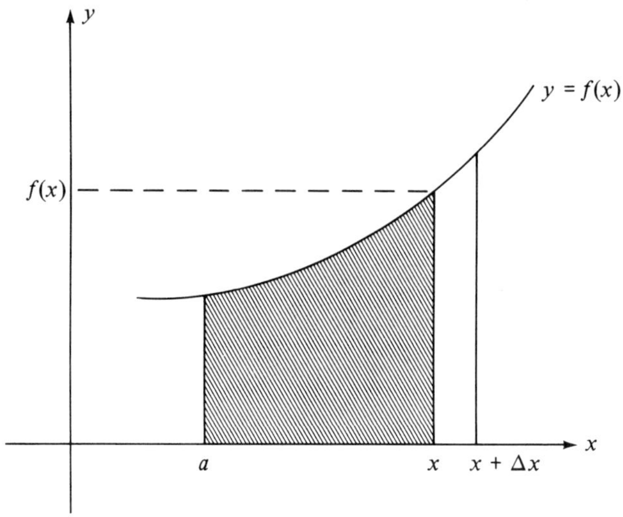

Fig. 5.1

Now, in the limit as $\Delta x \to 0$,

$$f(x) = \lim_{\Delta x \to 0} \left\{ \frac{A(x + \Delta x) - A(x)}{\Delta x} \right\}$$

$$= \frac{\mathrm{d}A}{\mathrm{d}x}$$

by the definition of a derivative (equation (1.2)). It follows that

$$\int f(x)\,\mathrm{d}x = A + C,$$

i.e.

$$A(x) = F(x) - C,$$

where $F(x)$ is a particular function obtained by integrating $f(x)$ $(F'(x) = f(x))$.

Since we must have $A(a) = 0$ then $0 = F(a) - C$ and the shaded area in Fig. 5.1 is given by

$$A(x) = F(x) - F(a).$$

It follows that for $f(x) \geqslant 0$ and $b > a$ the *area under the curve* $y = f(x)$ between $x = a$ and $x = b$ is given by $F(b) - F(a)$, where $F(x)$ is any function such that $F'(x) = f(x)$. We use the symbol $[F(x)]_a^b$ to denote $F(b) - F(a)$.

The process of integrating $f(x)$ to obtain $F(x)$ and then evaluating $F(b) - F(a)$ is described by the notation

$$\int_a^b f(x)\,\mathrm{d}x.$$

This is called the **definite integral** of $f(x)$ over the closed interval between a and b; a and b are called the **limits of integration**. (The restrictions $f(x) \geqslant 0$ and $b > a$ are no longer necessary for this definition. However, $f(x)$ must be *continuous* in the closed interval between a and b (see problem 6 in section 1.2).)

Summarizing,

$$\int_a^b f(x)\,\mathrm{d}x = [F(x)]_a^b = F(b) - F(a), \tag{5.1}$$

where $F(x)$ is any *indefinite integral* of $f(x)$.

For example,

$$\begin{aligned}\int_2^3 (3x^2 + 1)\,\mathrm{d}x &= [x^3 + x]_2^3 \\ &= (3^3 + 3) - (2^3 + 2) \\ &= 20.\end{aligned}$$

Note that the answer does not depend on x. We would still get 20 by evaluating

$$\int_2^3 (3t^2 + 1)\,\mathrm{d}t$$

or

$$\int_2^3 (3u^2 + 1)\,\mathrm{d}u.$$

The variable x in

$$\int_a^b f(x)\,\mathrm{d}x$$

is called a **dummy variable** (any letter will do). The value of the definite integral

$$\int_a^b f(x)\,\mathrm{d}x$$

depends on the *function* f and the integration limits a and b.

The following properties of a definite integral are easily proved from the definition:

$$\int_b^a f(x)\,\mathrm{d}x = -\int_a^b f(x)\,\mathrm{d}x, \tag{5.2}$$

$$\int_a^c f(x)\,\mathrm{d}x + \int_c^b f(x)\,\mathrm{d}x = \int_a^b f(x)\,\mathrm{d}x, \tag{5.3}$$

$$\frac{\mathrm{d}}{\mathrm{d}t}\left\{\int_a^t f(x)\,\mathrm{d}x\right\} = f(t). \tag{5.4}$$

In this section, we have approached the definition of a definite integral by consideration of the area under a curve. Clearly the calculation of such areas is carried out by the evaluation of the corresponding definite integral. It is important to understand that the calculation of areas is *only one* of the applications of definite integrals. In section 5.3, they are applied to the calculation of *volumes* and, in Chapter 11, applications include the calculation of *arc lengths, surface area, position of centroid, second moment of area* and *moment of intertia*.

The remainder of this section concentrations on the *evaluation* of certain definite integrals.

Example 5.1

Evaluate the following definite integrals.

(i) $\displaystyle\int_2^4 \left(x^2 + \frac{1}{x^2}\right) \mathrm{d}x.$

(ii) $\displaystyle\int_0^{2\pi} \cos^2 \theta \,\mathrm{d}\theta.$

(iii) $\displaystyle\int_0^2 x(x-1)(x-2)\,\mathrm{d}x.$

(iv) $\displaystyle\int_0^3 \sqrt{x+1}\,\mathrm{d}x.$

(i)
$$\int_2^4 \left(x^2 + \frac{1}{x^2}\right) \mathrm{d}x = \left[\frac{x^3}{3} - \frac{1}{x}\right]_2^4$$
$$= \left(\frac{64}{3} - \frac{1}{4}\right) - \left(\frac{8}{3} - \frac{1}{2}\right)$$
$$= \frac{227}{12}.$$

(ii)
$$\int_0^{2\pi} \cos^2 \theta \,\mathrm{d}\theta = \int_0^{2\pi} \tfrac{1}{2}(1 + \cos 2\theta)\,\mathrm{d}\theta$$
$$= \tfrac{1}{2}\left[\theta + \tfrac{1}{2}\sin 2\theta\right]_0^{2\pi}$$
$$= \tfrac{1}{2}\left\{(2\pi + \tfrac{1}{2}\sin 4\pi) - (0 + \tfrac{1}{2}\sin 0)\right\}$$
$$= \pi.$$

(iii)
$$\int_0^2 x(x-1)(x-2)\,\mathrm{d}x = \int_0^2 (x^3 - 3x^2 + 2x)\,\mathrm{d}x$$
$$= \left[\frac{x^4}{4} - x^3 + x^2\right]_0^2$$

$$= (4 - 8 + 4) - 0$$

$$= 0.$$

(iv) Substitute $u = x + 1$, $du = dx$.

$$\int \sqrt{x+1}\, dx = \int u^{1/2}\, du$$

$$= \tfrac{2}{3} u^{3/2} + C$$

$$= \tfrac{2}{3} (x+1)^{3/2} + C.$$

Therefore,

$$\int_0^3 \sqrt{x+1}\, dx = [\tfrac{2}{3}(x+1)^{3/2}]_0^3,$$ Constant of integration not required.

$$= \tfrac{2}{3}\{4^{3/2} - 1^{3/2}\}$$

$$= \tfrac{14}{3}.$$

The working for (iv) may be shortened by simply changing the limits of integration so that they correspond to the new variable u. It is then unnecessary to express the indefinite integral in terms of the original variable, i.e. put $u = x + 1$, $du = dx$. When $x = 0, u = 1$; when $x = 3, u = 4$. Therefore,

$$\int_0^3 \sqrt{x+1}\, dx = \int_1^4 u^{1/2}\, du$$

$$= [\tfrac{2}{3} u^{3/2}]_1^4$$

$$= \tfrac{2}{3}(4^{3/2} - 1^{3/2})$$

$$= \tfrac{14}{3}.$$

Example 5.2

Evaluate the following integrals.

(i) $\displaystyle\int_0^2 t\sqrt{5 + t^2}\, dt.$

(ii) $\displaystyle\int_0^{\pi/2} (1 + \cos^2 2x) \sin 2x\, dx.$

(i) Substitute $u = 5 + t^2$, $du = 2t\, dt$ and $t\, dt = \frac{1}{2}\, du$. When $t = 0$, $u = 5$; when $t = 2$, $u = 9$.

$$\int_0^2 t\sqrt{5+t^2}\,\mathrm{d}t = \int_5^9 u^{1/2} \times \tfrac{1}{2}\,\mathrm{d}u$$
$$= \tfrac{1}{2}\left[\tfrac{2}{3}u^{3/2}\right]_5^9$$
$$= \tfrac{1}{3}(9^{3/2} - 5^{3/2})$$
$$= 9 - \frac{5\sqrt{5}}{3}.$$

(ii) Substitute $u = \cos 2x$, $\mathrm{d}u = -2\sin 2x\,\mathrm{d}x$ and $\sin 2x\,\mathrm{d}x = -\tfrac{1}{2}\,\mathrm{d}u$. When $x = 0$, $u = 1$; when $x = \pi/2, u = -1$.

$$\int_0^{\pi/2} (1 + \cos^2 2x)\sin 2x\,\mathrm{d}x = \int_1^{-1} (1+u^2)(-\tfrac{1}{2}\,\mathrm{d}u)$$
$$= -\tfrac{1}{2}\int_1^{-1} (1+u^2)\,\mathrm{d}u$$

From equation (5.2),

$$-\tfrac{1}{2}\int_1^{-1} (1+u^2)\,\mathrm{d}u = \tfrac{1}{2}\int_{-1}^{1} (1+u^2)\,\mathrm{d}u$$
$$= \tfrac{1}{2}\left[u + \frac{u^3}{3}\right]_{-1}^{1}$$
$$= \tfrac{1}{2}\{\tfrac{4}{3} - (-\tfrac{4}{3})\}$$
$$= \tfrac{4}{3}.$$

Problems

1. Evaluate the following definite integrals.

(i) $\displaystyle\int_0^{\pi/2} \sin 2x\,\mathrm{d}x.$

(ii) $\displaystyle\int_{-2}^{1} \frac{t^3+1}{t^2}\,\mathrm{d}t.$

(iii) $\displaystyle\int_{\pi/4}^{0} \sec x \tan x\,\mathrm{d}x.$

(iv) $\displaystyle\int_0^2 \frac{x^3}{(4+x^2)^{3/2}}\,\mathrm{d}x.$

(Hint: put $u = 4 + x^2$.)

(v) $\displaystyle\int_0^2 \frac{x}{\sqrt{2+x}}\,dx.$

(vi) $\displaystyle\int_0^{\pi/3} \sin^2\left(\frac{\pi}{3}-\theta\right)d\theta.$

2. The function $F(t)$ is defined by

$$F(t) = \int_1^t (1+u^2)^2\,du.$$

Verify that $F'(t) = (1+t^2)^2$.

Answers

1. (i) 1.
   (ii) -1.
   (iii) $1-\sqrt{2}$.
   (iv) $3\sqrt{2}-4$.
   (v) $\frac{8}{3}(\sqrt{2}-1)$.
   (vi) $\dfrac{\pi}{6}-\dfrac{\sqrt{3}}{8}$.

5.3 CALCULATION OF AREAS AND VOLUMES

We have already seen that the area under the curve $y = f(x)$, $f(x) \geqslant 0$, in the interval $a \leqslant x \leqslant b$ is given by the definite integral

$$\int_a^b f(x)\,dx.$$

We use this result to derive Theorem 5.1

Theorem 5.1 (Fundamental theorem of integral calculus)
Given that

$$a = x_0 < x_1 < x_2 < \ldots < x_{n-1} < x_n = b$$

and that

$$x_i - x_{i-1} = \Delta x = \frac{b-a}{n}$$

for $i = 1, 2, 3, \ldots, n$ (see Fig. 5.2), then

$$\int_a^b f(x)\,dx = \lim_{\Delta x\to 0}\left\{\sum_{i=1}^{n} f(x_i)\,\Delta x\right\}.$$

(Note that $\Delta x \to 0$ implies that $n \to \infty$.) (In less formal language the theorem simply states that a definite integral is the limit of a summation as each term in the summation tends to zero while the number of terms tends to infinity.)

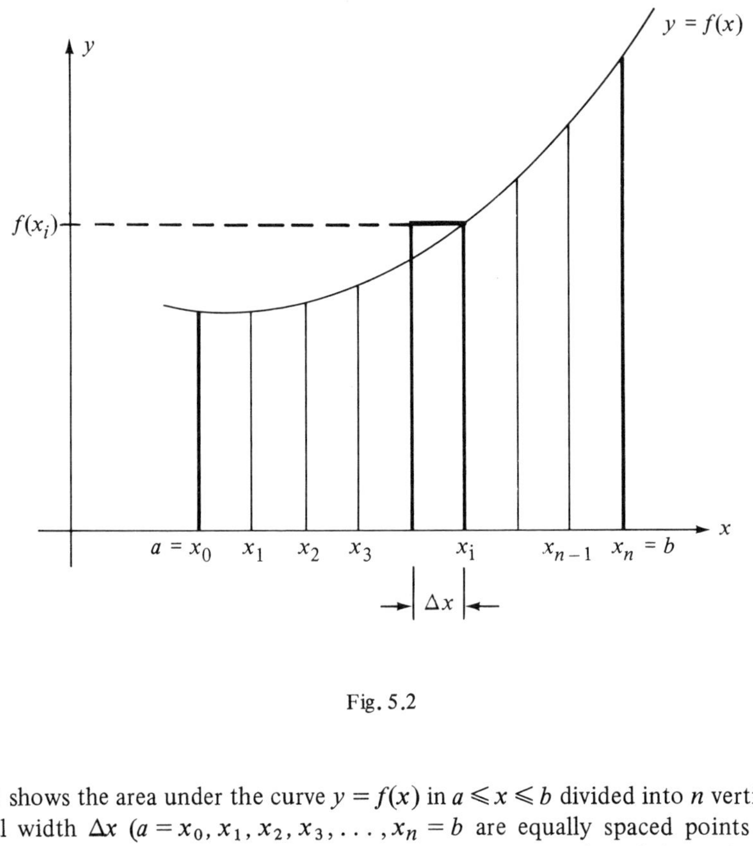

Fig. 5.2

Proof
Fig. 5.2 shows the area under the curve $y = f(x)$ in $a \leqslant x \leqslant b$ divided into n vertical strips of equal width Δx ($a = x_0, x_1, x_2, x_3, \ldots, x_n = b$ are equally spaced points on the x axis). Denoting this area by A, we note than an *approximate* value of A may be obtained by summing n rectangles; a typical rectangle of area $f(x_i)\ \Delta x$ is shown in Fig. 5.2. Therefore,

$$A \approx \sum_{i=1}^{n} f(x_i)\ \Delta x$$

and this approximation becomes more accurate if the number of strips is increased (Δx reduced).

We know from section 5.2 that

$$A = \int_a^b f(x)\ \mathrm{d}x$$

exactly. It follows that

$$\int_a^b f(x)\,dx = \lim_{\Delta x \to 0} \left\{ \sum_{i=1}^{n} f(x_i)\,\Delta x \right\}.$$

The fundamental result is used in many applications of integration, but for convenience the formal procedure of the theorem is relaxed by writing dx instead of Δx for the strip width and using the integral sign to denote summation. The theorem is applied in all examples (except for simple area calculations).

The volumes calculated in this section are all generated by the rotation of a given area about a given line. The area is divided into strips and one of the following methods is applied.

(i) *The method of discs* is used when the strips are drawn *perpendicular* to the axis of rotation.
(ii) *The method of shells* is used when the strips are drawn *parallel* to the axis of rotation.

The choice of method usually depends on the shape of the area. Examples 5.5–5.7 demonstrate the procedure. Theorem 5.1 is applied in each case. *It is not recommended* that formulae are remembered.

Example 5.3
Calculate the following areas.

(i) The area under the curve $y = 1 + x^2$ between $x = 1$ and $x = 3$.
(ii) The area between the x axis and the curve $y = (x + 1)(x - 4)$ in the interval $0 \leqslant x \leqslant 2$.
(iii) The area enclosed between the x axis and the curve $y = x(x - 1)(x - 2)$ in the interval $0 \leqslant x \leqslant 2$.
(iv) The area enclosed between the curves $y = x(2 - x)$ and $y = x/2$.

(i) A sketch of the required area is shown in Fig. 5.3.

$$\begin{aligned} \text{Area} &= \int_1^3 (1 + x^2)\,dx \\ &= \left[x + \frac{x^3}{3} \right]_1^3 \\ &= 12 - \tfrac{4}{3} \\ &= \tfrac{32}{3}\ \text{units}^2. \end{aligned}$$

(It is customary to write units^2 after the numerical result of an area calculation when the units are not specified. If the graph in Fig. 5.3 had been drawn with 1 unit = 1 cm on both the x axis and the y axis then the area would be $\frac{32}{3}$ cm^2.)

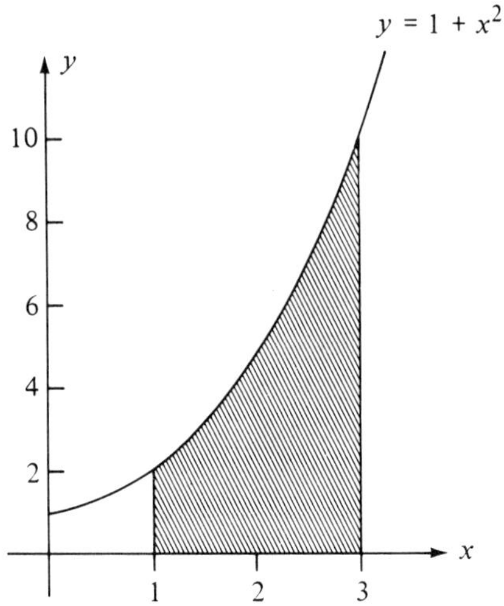

Fig. 5.3

(ii) Fig. 5.4 shows the required area, with a typical rectangular strip drawn. Since the function $y = (x+1)(x-4)$ is *negative* in $0 \leqslant x \leqslant 2$, then the numerical height of the rectangle is $-y$ and its area is $-y\,\mathrm{d}x$. The required area is obtained by summing all the rectangles and applying Theorem 5.1, i.e.

$$\begin{aligned}\text{area} &= \int_0^2 -y\,\mathrm{d}x\\ &= -\int_0^2 (x+1)(x-4)\,\mathrm{d}x\\ &= -\int_0^2 (x^2-3x-4)\,\mathrm{d}x\\ &= -\left[\frac{x^3}{3}-\frac{3x^2}{2}-4x\right]_0^2\\ &= \tfrac{34}{3}\ \text{units}^2.\end{aligned}$$

We note that, when $f(x) \leqslant 0$ and $a < b$, then

$$\int_a^b f(x)\,\mathrm{d}x$$

is *negative*. In this case the area between the curve $y = f(x)$ and the x axis in the interval $a \leqslant x \leqslant b$ is given by

$$-\int_a^b f(x)\,\mathrm{d}x.$$

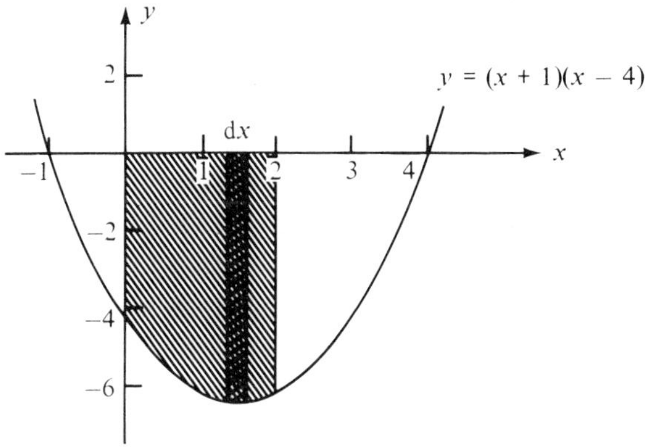

Fig. 5.4

(iii) The required area is shaded in Fig. 5.5.

$$\text{Area above } x \text{ axis} = \int_0^1 (x^3 - 3x^2 + 2x)\,dx$$

$$= \left[\frac{x^4}{4} - x^3 + x^2\right]_0^1$$

$$= \tfrac{1}{4} - 0$$

$$= \tfrac{1}{4}.$$

$$\text{Area below } x \text{ axis} = -\int_1^2 (x^3 - 3x^2 + 2x)\,dx$$

$$= -\left[\frac{x^4}{4} - x^3 + x^2\right]_1^2$$

$$= -\left\{(4 - 8 + 4) - (\tfrac{1}{4} - 1 + 1)\right\}$$

$$= \tfrac{1}{4}.$$

Therefore the total area $= \frac{1}{4} + \frac{1}{4} = \frac{1}{2}$ units^2. (Note that

$$\int_0^2 (x^3 - 3x^2 + 2x)\,dx = 0,$$

which illustrates the importance of drawing a rough sketch in *every* case.)

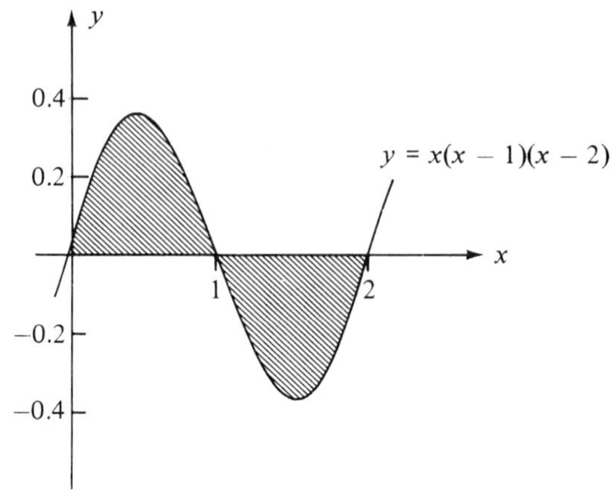

Fig. 5.5

(iv) The required area is shown in Fig. 5.6. The curves intersect where

$$x(2-x)=\frac{x}{2},$$

i.e. where

$$\frac{3x}{2}-x^2=0.$$

Thus, $x = 0$ and $x = \frac{3}{2}$. The area is divided into rectangular strips (a typical strip is shown in the figure) each of width dx. If we call the curves $y_T = x(2-x)$ and $y_B = x/2$ (y_T and y_B refer to *top* and *bottom* curves), then the area of the typical strip is $(y_T - y_B)\,dx$. Applying Theorem 5.1,

$$\begin{aligned}\text{required area} &= \int_0^{3/2} (y_T - y_B)\,dx\\ &= \int_0^{3/2} \left(\frac{3x}{2}-x^2\right) dx\\ &= \left[\frac{3x^2}{4}-\frac{x^3}{3}\right]_0^{3/2}\\ &= \left(\tfrac{27}{16}-\tfrac{9}{8}\right)-0\\ &= \tfrac{9}{16}\text{ units}^2.\end{aligned}$$

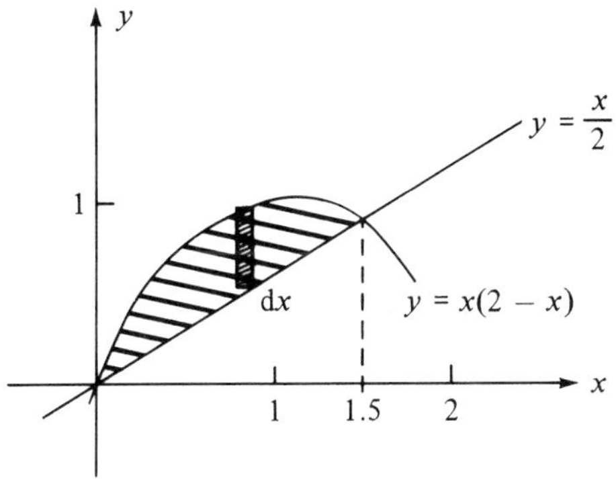

Fig. 5.6

Example 5.4
Determine the area enclosed between the curves $y^2 = x$ and $y = x - 2$.

The area is shown in Fig. 5.7; it is easily proved that the curves intersect at the points $(1, -1)$ and $(4, 2)$. To divide the area into vertical strips is not appropriate for this example since the 'bottom curve' has different equations for the intervals $0 \leqslant x \leqslant 1$ and $1 \leqslant x \leqslant 4$. This implies that *two* integrals are necessary which must be added to obtain the required area. This complication is avoided by dividing the area into horizontal strips (a typical strip is shown in the figure) each of width $\mathrm{d}y$. The area of this strip is $(x_R - x_L)\,\mathrm{d}y$ (x_R and x_L refer to *right* and *left* curves), and by Theorem 5.1 the required area is

$$\int_{-1}^{2} (x_R - x_L)\,\mathrm{d}y.$$

Now $x_R = y + 2, x_L = y^2$ and

$$\begin{aligned}\text{required area} &= \int_{-1}^{2} (y + 2 - y^2)\,\mathrm{d}y \\ &= \left[\frac{y^2}{2} + 2y - \frac{y^3}{3}\right]_{-1}^{2} \\ &= \tfrac{10}{3} - (-\tfrac{7}{6}) \\ &= \tfrac{9}{2}\ \text{units}^2.\end{aligned}$$

Example 5.5
The area under the curve $y = x^2 + 1$ between $x = 1$ and $x = 2$ is rotated by the x axis to generate a solid. Calculate the volume of this solid.

The volume, shown in Fig. 5.8, is divided into slices of equal width $\mathrm{d}x$. Each slice is approximately a thin circular cylinder or *disc* (a typical disc is shown in the figure). The volume of the typical disc is

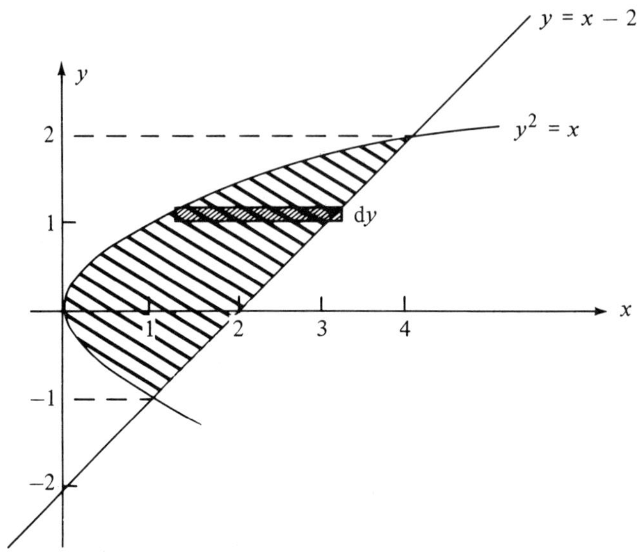

Fig. 5.7

$$\pi \times \text{radius}^2 \times \text{thickness} = \pi y^2 \text{ d}x.$$

The required volume is obtained by summing all the discs and applying Theorem 5.1, i.e.

$$\begin{aligned}
\text{volume} &= \int_1^2 \pi y^2 \text{ d}x \\
&= \pi \int_1^2 (x^2 + 1)^2 \text{ d}x \\
&= \pi \int_1^2 (x^4 + 2x^2 + 1) \text{ d}x \\
&= \pi \left[\frac{x^5}{5} + \frac{2x^3}{3} + x \right]_1^2 \\
&= \pi \left(\frac{206}{15} - \frac{28}{15} \right) \\
&= \frac{178\pi}{15} \text{ units}^3.
\end{aligned}$$

The method used in this example is the *method of discs*. Note that the discs are formed by the rotation of strips in the generating area which are *perpendicular* to the axis of rotation. This choice is appropriate since *all* such strips lie between the curves $y = x^2 + 1$ and $y = 0$.

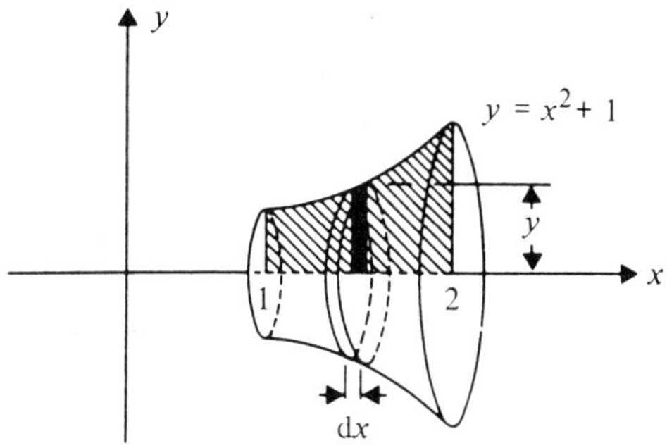

Fig. 5.8 The method of discs. The shaded area is rotated about the x axis. A typical strip sweeps out a thin disc.

Example 5.6

The area described in the previous example is rotated about the y axis. Find the volume of the solid generated.

As in the previous example, the generating area is divided into strips which are perpendicular to the x axis but this time *parallel* to the axis of rotation. The result is that each strip sweeps out a ring or thin cylindrical shell and the required volume consists of a number of concentric thin cylindrical shells. Fig. 5.9 shows the required volume together with a typical cylindrical shell. The volume of the typical shell is obtained by multiplying the strip area by the distance through which it travels as it rotates about the y axis. Hence the volume of the typical shell is

$$\text{strip area} \times 2\pi \times \text{radius} = y\,dx\,2\pi x.$$

The required volume is obtained by summing all the shells and applying Theorem 5.1, i.e.

$$\begin{aligned}
\text{volume} &= \int_1^2 2\pi xy\,dx \\
&= 2\pi \int_1^2 x(x^2+1)\,dx \\
&= 2\pi \int_1^2 (x^3+x)\,dx \\
&= 2\pi \left[\frac{x^4}{4} + \frac{x^2}{2}\right]_1^2 \\
&= 2\pi(6 - \tfrac{3}{4}) \\
&= \frac{21\pi}{2}\ \text{units}^3.
\end{aligned}$$

The method used in the example is the *method of shells*. The important feature of the method is that the generating area is divided into strips which are *parallel* to the axis of rotation.

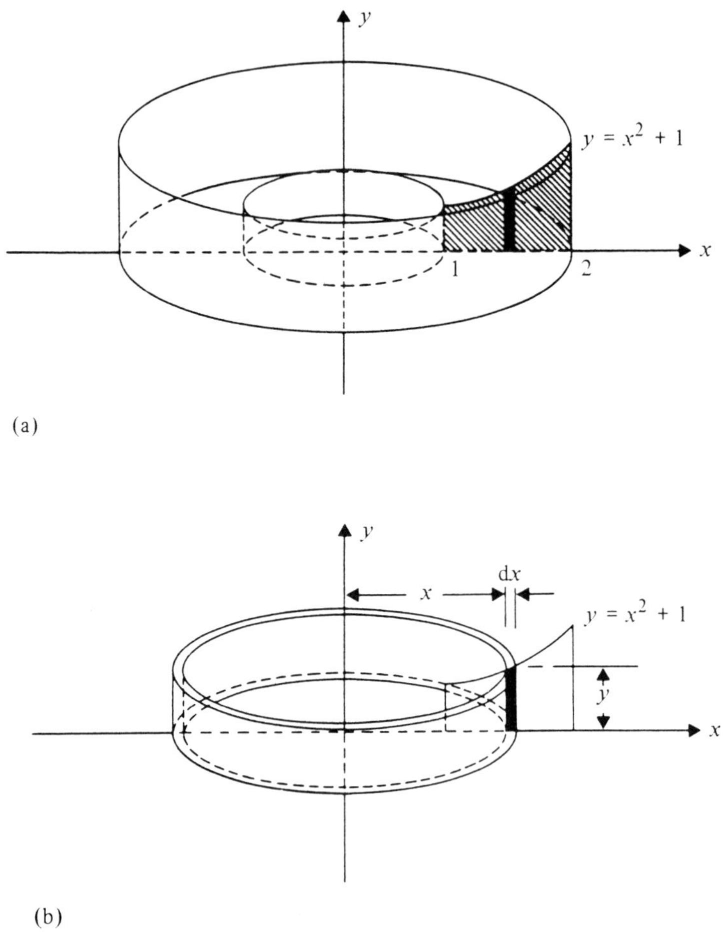

Fig. 5.9 The method of shells: (a) The shaded area is rotated about the y axis; (b) a typical strip sweeps out a thin cylindrical shell.

Example 5.7

A casting has the same shape as is swept out by the area shown in Fig. 5.10(a) when it is rotated about the line $y = 3$. Determine the volume of the casting.

Both methods (discs and shells) are demonstrated in this example so that the reader can make a comparison.

(i) **Method of discs**. The given area is divided into strips drawn perpendicular to the axis of rotation. The width of each strip, measured in the direction of the x axis, is denoted by $\mathrm{d}x$. Each strip, when rotated about the line $y = 3$, sweeps out an element shaped like a 'washer' (a washer is a disc with a smaller concentric disc removed from its centre). Fig. 5.10(b), (c) shows the required volume and a typical washer. The volume of the washer-shaped element is

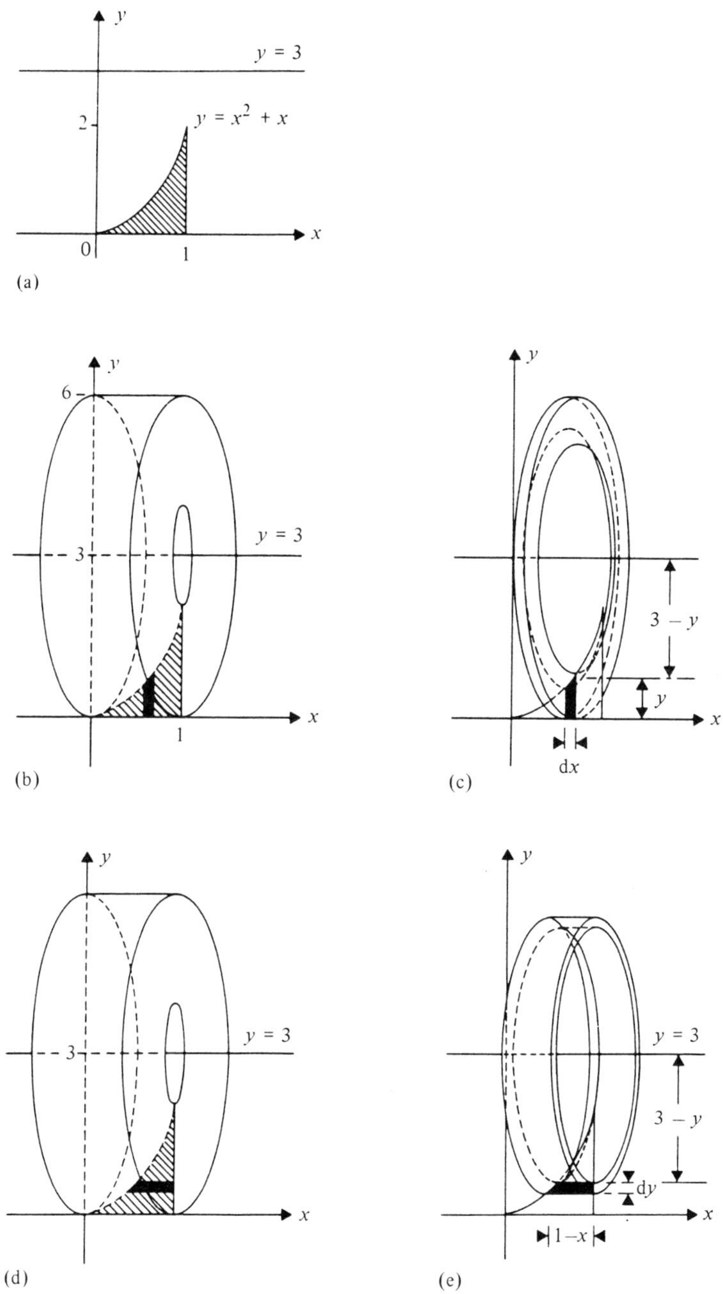

Fig. 5.10 (a) The problem and the solution (b), (c) by the method of discs and (d), (e) by the method of shells: (b) the shaded area is rotated about line $y = 3$; (c) a typical strip sweeps out a thin washer; (d) the shaded area is rotated about line $y = 3$; (e) a typical strip sweeps out a thin cylindrical shell.

$$\pi \times (\text{outer radius})^2 \times \text{thickness} - \pi \times (\text{inner radius})^2 \times \text{thickness}$$
$$= \pi \times 3^2 \,\mathrm{d}x - \pi(3-y)^2 \,\mathrm{d}x.$$

This is simply the difference of two discs. Applying Theorem 5.1,

$$\text{volume of casting} = \int_0^1 \pi\left\{3^2 - (3-y)^2\right\} \mathrm{d}x$$

$$= \pi \int_0^1 (6y - y^2)\,\mathrm{d}x$$

$$= \pi \int_0^1 (6x^2 + 6x - x^4 - 2x^3 - x^2)\,\mathrm{d}x$$

$$= \pi \left[\frac{5x^3}{3} + 3x^2 - \frac{x^5}{5} - \frac{x^4}{2}\right]_0^1$$

$$= \frac{119\pi}{30}\,\mathrm{m}^3.$$

(ii) **Method of shells.** The required area is now divided into strips drawn parallel to the axis of rotation. The width of each strip is now measured in the direction of the y axis and is denoted by $\mathrm{d}y$. Each strip, when rotated about the line $y = 3$, sweeps out a thin cylindrical shell; the required volume and a typical shell are shown in Fig. 5.10(d), (e). The volume of the typical shell is

$$2\pi \times \text{radius} \times \text{strip area} = 2\pi(3-y)(1-x)\,\mathrm{d}y$$
$$= 2\pi(3 - x - x^2)(1-x)(1+2x)\,\mathrm{d}x.$$

Applying Theorem 5.1,

$$\text{volume of casting} = 2\pi \int_0^1 (3 + 2x - 8x^2 + x^3 + 2x^4)\,\mathrm{d}x$$

$$= 2\pi \left[3x + x^2 - \frac{8x^3}{3} + \frac{x^4}{4} + \frac{2x^5}{5}\right]_0^1$$

$$= \frac{119\pi}{30}\,\mathrm{m}^3.$$

Example 5.8

The average car can decelerate at $0.8g$ on a dry level road. Find the total emergency stopping distance s, measured from the point where the driver first sees the danger, for a car travelling at 30 m/s. The reaction time of the driver is such that there is a delay of 0.75 s between the sighting of the danger and the application of the brakes. (Take $g = 9.81\ \mathrm{m/s^2}$.)

In section 4.5, we saw that distance travelled s could be obtained by *integrating* velocity v with respect to time, the constant of integration being evaluated from a given initial condition. More specifically we can express the distance travelled between the instants $t = t_1$ and $t = t_2$ by the *definite* integral

$$\int_{t_1}^{t_2} v \, dt$$

(assuming that $v \geqslant 0$).

In this chapter, we have seen the relationship between a definite integral and the area under a curve. It follows that distance travelled can be obtained by calculating the area under the v versus t curve.

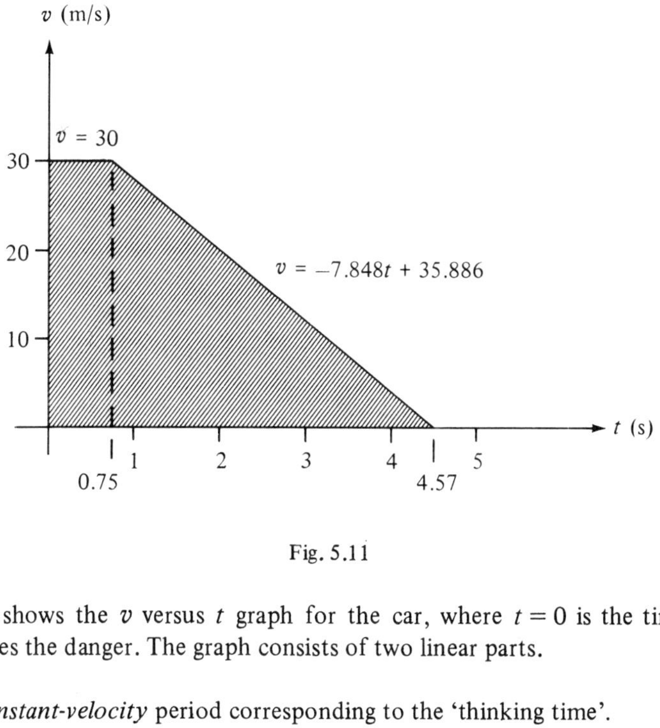

Fig. 5.11

Fig. 5.11 shows the v versus t graph for the car, where $t = 0$ is the time when the driver first sees the danger. The graph consists of two linear parts.

(i) The *constant-velocity* period corresponding to the 'thinking time'.
(ii) The *constant-deceleration* (0.8×9.81 m/s^2) during braking.

The equation for (ii) is $v - 30 = -0.8 \times 9.81(t - 0.75)$, (see section 0.6 (iii)), and the intersection of this straight line with the t axis occurs when $v = 0$, i.e.

$$t = \frac{30}{0.8 \times 9.81} + 0.75$$

$$\approx 4.57 \text{ s.}$$

The required area, shaded in Fig. 5.11, is simply

$$30 \times 0.75 + 0.5 \times 30(4.57 - 0.75) = 79.8$$

This area corresponds to the distance travelled by the car and the units are *metres* since, in calculating the area, we are effectively multiplying velocity (m/s) by time (s). Hence the stopping distance for the car is approximately 79.8 m.

Problems

1. Calculate the following areas.
   (i) The area under $y = \sqrt{2x + 1}$ between $x = 0$ and $x = 4$.
   (ii) The area bounded by

   $$y = \frac{1}{(2x + 1)^2},$$

   the x axis and the lines $x = 1$ and $x = 2$.
   (iii) The area enclosed between the curve $y = x^2 - 4$ and the x axis.
   (iv) The area enclosed between the x axis and one arch of the curve $y = \cos 3x$.
   (v) The area enclosed by the curves $y = x^2$ and $y = 2x$.
   (vi) The area enclosed by the y axis and the curve $x = y^2 - y^3$.
   (vii) The area enclosed between the curves $y = 2 - x^2$ and $y = -x$.
   (viii) The area bounded by the curve $x = 3y - y^2$ and the line $x + y = 3$.

2. The area enclosed between the curve $y = x^2$ and the line $y = 4$ is divided into equal portions by the line $y = c$. Find c.

3. Calculate the following volumes.
   (i) The volume generated when the area enclosed by $x + y = 2$, $x = 0$, $y = 0$, is rotated about the x axis.
   (ii) The volume generated when the area enclosed by $y = x - x^2$ and the x axis is rotated about the x axis.
   (iii) The volume generated when the area enclosed between $y = 3x - x^2$ and $y = x$ is rotated about the x axis.
   (iv) The volume generated when the area enclosed between $x = 2y - y^2$ and the y axis is rotated about the x axis.
   (v) The volume generated when the area enclosed by $y = \sqrt{x}$, $y = 0$, $x = 1$ and $x = 9$ is rotated about the y axis.
   (vi) The volume generated when the area enclosed by $x^2 + y^2 = 16$, $x = \sqrt{7}$ and $x = \sqrt{12}$, is rotated about the y axis.
   (vii) The volume generated when the area enclosed by $x = 2y - y^2$ and $x = 0$ is rotated about the y axis.

4. Find the volume generated when the area between $y = x^2$ and $y = x$ is rotated
   (i) about the x axis and
   (ii) about the y axis.

5. The area enclosed by $y = x^2$ and $y = 4$ generates various solids when rotated about the following axes.
   (i) The x axis.
   (ii) The y axis.
   (iii) The line $y = 4$.
   (iv) The line $y = -1$.
   (v) The line $x = 2$.
   Find the volume of each solid.

6. The shape of a cooling tower is generated by the rotation about the line $x = -6$ of the area enclosed by $y^2 = 100x$, $x = -6$, $y = 0$, $y = -20$, (Fig. 5.12). Calculate the volume enclosed by the cooling tower.

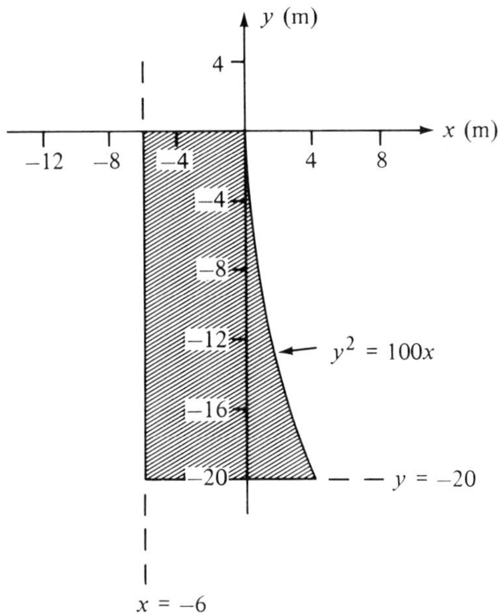

Fig. 5.12

7. Use integration to confirm the following results.
   (i) Volume of sphere $= \frac{4}{3}\pi \times \text{radius}^3$.
   (ii) Volume of cone $= \frac{1}{3}\pi \times (\text{base radius})^2 \times \text{height}$.

8. Sketch the curve

$$y = \sqrt{a^2 - x^2}.$$

Hence evaluate

$$\int_{-a}^{a} \sqrt{a^2 - x^2}\, dx.$$

A torus is formed by rotating the area enclosed by the circle $x^2 + y^2 = a^2$ about the line $x = R$. Show that the volume of the torus is $2\pi R \times$ area of the circle. (Assume that $R > a$.)

9. Show that the volume V of liquid in a hemispherical bowl of radius a is given by

$$V = \pi h^2 \left(a - \frac{h}{3} \right),$$

where h is the depth of liquid. If water runs into a hemispherical bowl of radius 1 m at a rate of 0.05 $\mathrm{m^3/s}$, how fast is the water level rising when the depth of water is 0.5 m? Use

$$\frac{\mathrm{d}V}{\mathrm{d}t} = \frac{\mathrm{d}V}{\mathrm{d}h}\frac{\mathrm{d}h}{\mathrm{d}t}.$$

10. A train accelerates from rest at the rate of 3 $\mathrm{m/s^2}$ until it reaches its maximum permissible speed of 45 m/s. After running at this speed for a certain time, the train decelerates at a rate of 2 $\mathrm{m/s^2}$ until it comes to a stop. If the train has travelled a total distance of 4 km, find the time required for the train journey correct to the nearest second.

Answers

1. (i) $\frac{26}{3}$.
   (ii) $\frac{1}{15}$.
   (iii) $\frac{32}{3}$.
   (iv) $\frac{2}{3}$.
   (v) $\frac{4}{3}$.
   (vi) $\frac{1}{12}$.
   (vii) $\frac{9}{2}$.
   (viii) $\frac{4}{3}$.
2. $\sqrt[3]{16}$.
3. (i) $\dfrac{8\pi}{3}$.
   (ii) $\dfrac{\pi}{30}$.
   (iii) $\dfrac{56\pi}{15}$.
   (iv) $\dfrac{8\pi}{3}$.

(v) $\dfrac{968\pi}{5}$.

(vi) $\dfrac{76\pi}{3}$.

(vii) $\dfrac{16\pi}{15}$.

4. (i) $\dfrac{2\pi}{15}$.

(ii) $\dfrac{\pi}{6}$.

5. (i) $\dfrac{256\pi}{5}$.

(ii) 8π.

(iii) $\dfrac{512\pi}{15}$.

(iv) $\dfrac{1088\pi}{15}$.

(v) $\dfrac{128\pi}{3}$.

6. 1104π m^3.

8. $\dfrac{\pi a^2}{2}$.

9. $\dfrac{1}{15\pi}$ m/s.

10. 108 s.

Further Problems for Chapter 5

1. Evaluate

$$\int_1^4 \frac{dx}{\sqrt{x}}.$$

(Note that

$$\int \frac{dx}{\sqrt{x}}$$

and

$$\int \frac{1}{\sqrt{x}} \, dx$$

are identical.)

2. Evaluate

$$\int_0^{\pi/2} \sin 3y \, dy.$$

3. Given that

$$\int_0^{0.5} \frac{dx}{\sqrt{1-x^2}} = \frac{\pi}{6},$$

evaluate

$$\int_0^{0.5} \frac{dt}{\sqrt{1-t^2}} .$$

4. Use the substitution $x = \sin \theta$ to show that

$$\int_0^{0.5} \frac{dx}{\sqrt{1-x^2}} = \int_0^{\pi/6} d\theta.$$

Hence confirm the result given in the previous example.

5. For a certain function $f(x)$ it is given that

$$\int_0^3 f(x) \, dx = 7,$$

$$\int_0^{10} f(x) \, dx = 4.$$

Evaluate the following.

(i) $\int_3^{10} f(x) \, dx.$

(ii) $\int_{10}^3 f(x) \, dx.$

6. The function $F(t)$ is defined by

$$F(t) = \int_0^t 3^x \, dx.$$

Write down the following.
(i) $F(0)$.
(ii) $F'(2)$.
(Do *not* attempt to evaluate the integral.)

7. The graph of a certain function $y = f(x)$ is symmetrical about the y axis. By considering the area under the curve $y = f(x)$ show that

$$\int_{-k}^{k} f(x)\,\mathrm{d}x = 2\int_{0}^{k} f(x)\,\mathrm{d}x.$$

Is this result true for *all* functions $f(x)$? If your answer is *no*, provide an example which does *not* satisfy the result.

8. Calculate the area enclosed by the curve $y = \sin x$ and the x axis in the interval $0 \leqslant x \leqslant \pi$.

9. Calculate the area enclosed between the curve $y = x^2 - 4$ and the x axis.

10. Determine the area bounded by the curve $y = x^2 - 4$ and the line $y = 5$.

11. The area under the curve $y = \sqrt{x}$ in the interval $0 \leqslant x \leqslant 1$ is rotated about the x axis to form a solid. Calculate the volume of the solid.

12. Calculate the volume generated when the area defined in the previous example is rotated about the y axis.

13. The rectangle bounded by the lines $x = 0$, $x = 2$, $y = 0$ and $y = 1$ is rotated about the line $y = -3$. Calculate the volume generated.

14. A car travelling at constant velocity 15 m/s in a straight line passes the point A. It continues at constant velocity until 10 s later when the brakes are applied resulting in a constant deceleration of 5 m/s^2. If the car comes to rest at B, calculate AB.

Answers

1. 2.
2. $\frac{1}{3}$.
3. $\dfrac{\pi}{6}$.
5. (i) -3,
   (ii) 3.
6. (i) 0.
   (ii) 9.

7. No, e.g. $\int_{-1}^{1} (4x + 3x^2)\,dx = 2$, but $2\int_{0}^{1} (4x + 3x^2)\,dx = 6$.

8. 2.
9. $\frac{32}{3}$.
10. 36.
11. $\frac{\pi}{2}$.
12. $\frac{4\pi}{5}$.
13. 14π.
14. 172.5 m.

6

Stationary Points and Points of Inflexion

6.1 INTRODUCTION

In Chapter 3, we considered briefly the significance of the *signs* of $f'(x)$ and $f''(x)$ in relation to the curve $y = f(x)$. In this chapter, detailed consideration is given to *stationary points* (points where $f'(x) = 0$) and *points of inflexion* (points where $f''(x)$ *changes sign*). Further examples of curve sketching are considered and, in section 6.4, applications to optimization problems are presented.

6.2 STATIONARY POINTS

For a *differentiable* function $y = f(x)$ (i.e. $f'(x)$ exists), we have the following definitions.

(i) If $f'(a) > 0$, $f(x)$ is said to be **increasing** at the point $x = a$ (a small increase in x results in an *increase* in y since $\mathrm{d}y/\mathrm{d}x > 0$).
(ii) If $f'(a) < 0$, $f(x)$ is said to be **decreasing** at the point $x = a$ (a small increase in x results in a *decrease* in y since $\mathrm{d}y/\mathrm{d}x < 0$).
(iii) If $f'(a) = 0$, $f(x)$ is said to be **stationary** at the point $x = a$. The point with co-ordinates $(a, f(a))$ is called a **stationary point** of the curve $y = f(x)$.

There are *three types* of stationary points: a **maximum point**, a **minimum point** and a **horizontal point of inflexion** (these are illustrated in Fig. 6.1). In all cases the curve has zero gradient, i.e. the tangent line at a stationary point is horizontal.

The words *maximum* and *minimum* are in common use and apply relative to a neighbourhood of the stationary point. The value of $f(x)$ at a maximum (or minimum) stationary point is not necessarily the greatest (or least) value of the function (see section 6.3).

It is important to understand that a *horizontal* point of inflexion is a *special type* of point of inflexion. General points of inflexion are considered in section 6.5.

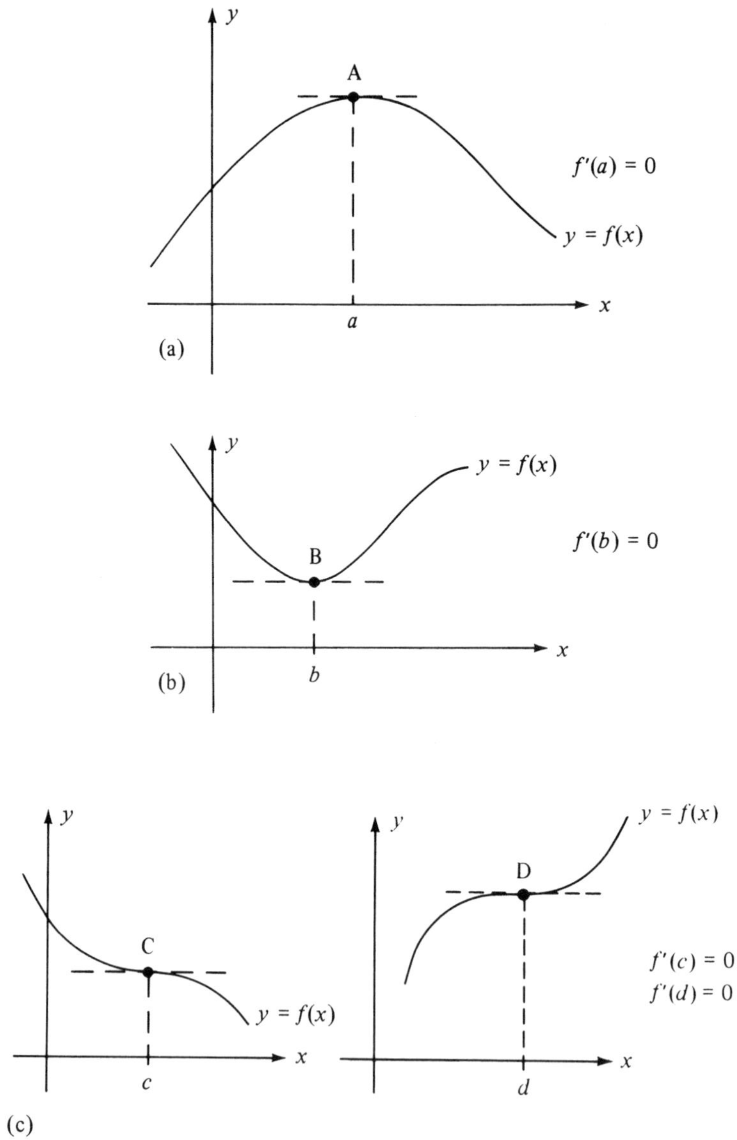

Fig. 6.1 Types of stationary point: (a) A is a maximum point; (b) B is a minimum point; (c) C and D are horizontal points of inflexion.

To obtain the x coordinates of all stationary points for a given curve $y = f(x)$, it is sufficient to solve the equation $f'(x) = 0$. For each solution, we may determine the *type* of stationary point by applying either Theorem 6.1 or Theorem 6.2. Theorem 6.1 considers the sign of $f'(x)$ *just before and just after* the stationary point; Theorem 6.2 evaluates $f''(x)$ *at* the stationary point in an attempt to determine concavity.

Theorem 6.1 (The first derivative test for differentiable functions $y = f(x)$)

(i) Given that $f'(a) = 0$, then there is a *maximum* stationary point at $x = a$ if and only if $f'(x)$ changes from positive to negative as x increases through the value $x = a$ (i.e. $f(x)$ is increasing just before $x = a$ and decreasing first after $x = a$; we say that $f'(x)$ 'goes' $+\,0\,-$ (see Fig. 6.1(a)).

(ii) Given that $f'(b) = 0$, then there is a *minimum* stationary point at $x = b$ if and only if $f'(x)$ changes sign from negative to positive as x increases through the value $x = b$ (we say that $f'(x)$ 'goes' $-\,0\,+$ (see Fig. 6.1(b)).

(iii) Given that $f'(c) = 0$, then the stationary point at $x = c$ is a *horizontal point of inflexion* if and only if $f'(x)$ *does not change sign* as x increases through the value $x = c$ (we say that $f'(x)$ 'goes' $+\,0\,+$ or $-\,0\,-$ (see Fig. 6.1(c)).

Summarizing Theorem 6.1,

$$f'(x) \text{ goes } +\,0\,- \Leftrightarrow \text{maximum stationary point,}$$

$$f'(x) \text{ goes } -\,0\,+ \Leftrightarrow \text{minimum stationary point,}$$

$$f'(x) \text{ goes } -\,0\,- \text{ or } +\,0\,+ \Leftrightarrow \text{horizontal inflexion.}$$

(The symbol $\Leftrightarrow$ means implies and is implied by, i.e. the statement is *reversible*.)

Theorem 6.2 (The second-derivative test for twice-differentiable functions $y = f(x)$)

(i) If $f'(a) = 0$ and $f''(a) < 0$, then there is a *maximum* stationary point at $x = a$. (This follows immediately from the discussion on concavity in section 3.2. The negative second derivative implies a *concave-down* curve and since the first derivative is zero we have a 'concave-down stationary point'.)

(ii) If $f'(b) = 0$ and $f''(b) > 0$, then there is a *minimum* stationary point at $x = b$ (a concave-up stationary point).

(Note that, if $f''(x) = 0$ at a stationary point, then *no conclusion* can be drawn about the type of point. Any type is possible and Theorem 6.1 should be used (see Example 6.1).)

Summarizing Theorem 6.2

$$f'(x) = 0,\ f''(x) < 0 \Rightarrow \text{maximum stationary point,}$$

$$f'(x) = 0,\ f''(x) > 0 \Rightarrow \text{minimum stationary point.}$$

It is important to understand that the above statements are *not reversible*. For example, $f''(x)$ is *not necessarily negative* at a maximum stationary point; it *may be zero*. The conditions (i) and (ii) in Theorem 6.2 are *sufficient conditions*; they are not *necessary conditions*. Theorem 6.1 is stronger, its conditions are necessary *and* sufficient.

Theorem 6.2 should only be used in situations where the second derivative is easy to obtain. However, in cases where $f''(x) = 0$ at the stationary point, Theorem 6.2 *does not apply* and it will be necessary to use Theorem 6.1.

Example 6.1

Confirm that each of the following curves has just one stationary point. Obtain its co-ordinates in each case and, using Theorem 6.1, determine the type of stationary point. Show that Theorem 6.2 does not apply for any of the curves.

(i) $y = x^4 + 1$.
(ii) $y = 9 - 3x + 3x^2 - x^3$.
(iii) $y = 5 - (2x + 3)^6$.

(i) $y = x^4 + 1$.

$$\frac{dy}{dx} = 4x^3.$$

To obtain stationary points, we solve the equation $dy/dx = 0$, i.e. $4x^3 = 0$, which has just one solution $x = 0$. Therefore the stationary point is at (0, 1). Applying Theorem 6.1, we note that when x is just before $x = 0$ (i.e. just negative) then $dy/dx = 4x^3$ is *negative*. When x is just after $x = 0$ (i.e. just positive) then $dy/dx = 4x^3$ is *positive*. Therefore, dy/dx goes $-0+$ and (0, 1) is a *minimum* stationary point.

(ii) $y = 9 - 3x + 3x^2 - x^3$.

$$\frac{dy}{dx} = -3 + 6x - 3x^2$$

$$= -3(x - 1)^2.$$

Clearly the only solution of $dy/dx = 0$ is at $x = 1$ and at all other points $dy/dx < 0$. Therefore, dy/dx goes $-0-$ as x increases through $x = 1$ and $(1, 8)$ is a *horizontal inflexion.*

(iii) $y = 5 - (2x + 3)^6$.

$$\frac{dy}{dx} = -12(2x + 3)^5.$$

The positions of stationary points are given by solutions of $-12(2x + 3)^5 = 0$, i.e. there is just one stationary point at $(-1.5, 5)$. The term $(2x + 3)^5$ changes from negative to positive as x increases through $x = -1.5$ and therefore $dy/dx = -12(2x + 5)^5$ goes $+0-$. Hence $(-1.5, 5)$ is a *maximum* stationary point.

Theorem 6.1 has been used in each of the above examples. Any attempt to use Theorem 6.2 would have been unsuccessful since in each case $d^2y/dx^2 = 0$ at the stationary point. In (i), $d^2y/dx^2 = 12x^2 = 0$ at (0, 1), in (ii), $d^2y/dx^2 = -6(x-1) = 0$ at (1, 8) and, in (iii), $d^2y/dx^2 = -120(2x+3)^4 = 0$ at (−1.5, 5).

We note again that *any* type of stationary point is possible when both dy/dx and d^2y/dx^2 are zero.

Example 6.2
Find the coordinates of the stationary points of the curve $y = x^4(6 - x^2)$ and determine the nature of each.

$$y = x^4(6 - x^2) = 6x^4 - x^6.$$

$$\frac{dy}{dx} = 24x^3 - 6x^5 = 6x^3(2-x)(2+x).$$

For this simple example, d^2y/dx^2 is easily obtained for use in Theorem 6.2, the second-derivative test:

$$\frac{d^2y}{dx^2} = 72x^2 - 30x^4.$$

$$\frac{dy}{dx} = 0$$

when $x = 2, x = -2, x = 0$.

Therefore, there are three stationary points (2, 32), (−2, 32) and (0, 0).

At (2, 32), $dy/dx = 0$ and $d^2y/dx^2 = -192 < 0$. Therefore, applying Theorem 6.2, (2, 32) is a *maximum* stationary point.

At (−2, 32), $dy/dx = 0$ and $d^2y/dx^2 = -192 < 0$. Therefore, applying Theorem 6.2, (−2, 32) is also a *maximum* stationary point.

At (0, 0), $dy/dx = 0$ and $d^2y/dx^2 = 0$. Theorem 6.2 does not apply and we use Theorem 6.1, the first-derivative test. As x increases through $x = 0$, $dy/dx = 6x^3(4 - x^2)$ goes $- \; 0 \; +$ and therefore (0, 0) is a *minimum* stationary point.

A rough sketch of the curve is shown in Fig. 6.2. Such a sketch is often useful for the detection of errors (e.g. it would be impossible to sketch a *continuous* curve with adjacent maximum stationary points).

Example 6.3
Obtain the stationary points of the curve

$$y = \frac{(x-3)^2}{(x-2)^2}.$$

Sketch the curve.

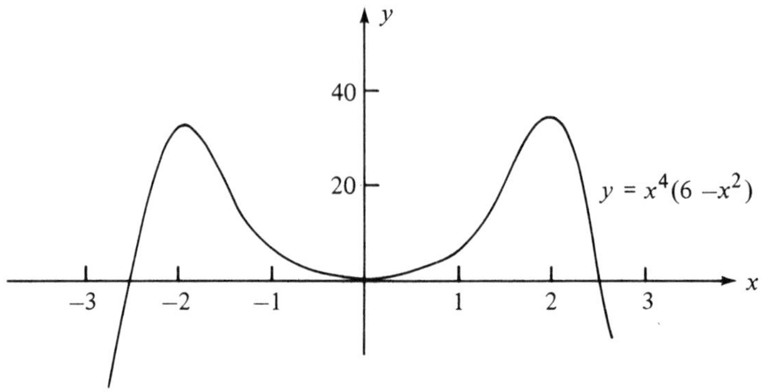

Fig. 6.2

Clearly, $x \neq 2$ ($x = 2$ is excluded from the domain of the function).

$$\frac{\mathrm{d}y}{\mathrm{d}x} = \frac{(x-2)^2 2(x-3) - (x-3)^2 2(x-2)}{(x-2)^4}$$

$$= \frac{2(x-2)(x-3)\{(x-2)-(x-3)\}}{(x-2)^4}$$

$$= \frac{2(x-3)}{(x-2)^3}.$$

Determination of the second derivative is not immediate and it is preferable to use Theorem 6.1, the first-derivative test.

$\mathrm{d}y/\mathrm{d}x = 0$ when $x = 3$ and $(3, 0)$ is the only stationary point. As x increases through the value $x = 3$, $\mathrm{d}y/\mathrm{d}x$ goes $-0+$ and, applying Theorem 6.1, $(3, 0)$ is a *minimum* stationary point.

The curve-sketching procedure is carried out as described in section 0.3. The sketch is shown in Fig. 6.3.

(a) **Restrictions.** $x \neq 2, y \geqslant 0$.
(b) **Axis intersections.** $(3, 0)$, $(0, \frac{9}{4})$.
(c) **Symmetry**. None.
(d) **Stationary points.** $(3, 0)$, minimum.
(e) **Asymptotes.**
  (i) $x = 2$; as $x \to 2^-$, $y \to \infty$ and, as $x \to 2^+$, $y \to \infty$.
  (ii) $y = 1$; as $x \to \infty$, $y \to 1$ and, as $x \to -\infty$, $y \to 1$.

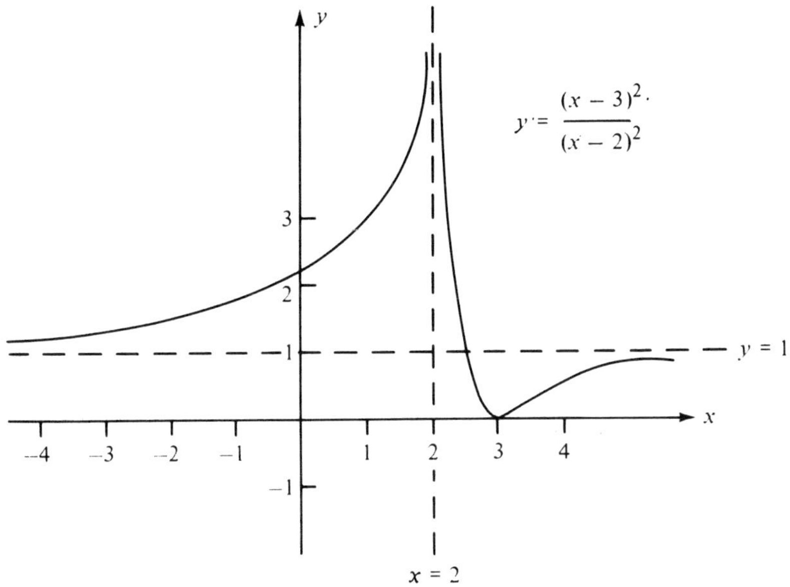

Fig. 6.3

Problems

1. (i) Show that the following functions are increasing for all values of x.
   (a) $y = 6x^3 + 3x + 2$.
   (b) $y = 3x + \cos 2x$.
   (ii) Find the range of values of x for which the function

$$y = x^3 - x^2 - x + 5$$

is decreasing. (Hint: sketch the quadratic graph $\mathrm{d}y/\mathrm{d}x$ versus x.)

2. Find the stationary points for the following functions and determine their nature. In each case make a rough sketch of the curve.
   (i) $y = x^2 - x + 1$.
   (ii) $y = 2x^3 - 3x^2 + 3$.
   (iii) $y = x^4 + 4x^3$.
   (iv) $y = x^4 - 8x^2 + 16$.
   (v) $y = 3x^5 - 5x^3 + 10$.
   (vi) $y = (x-1)^4(x+2)^3$.
   (vii) $y = x\sqrt{x+1}$, $x \geqslant -1$.
   (viii) $y = \dfrac{(2+x)^3}{(1+x)^2}$, $x \neq -1$.

3. (i) Prove that the function

$$y = \sin^3 x \cos x$$

has a maximum at $x = \pi/3$.

(ii) Prove that the function

$$y = \frac{\sin x}{1 + \tan x}$$

has a maximum at $x = \pi/4$.

4. Show that the curve $y = x^5 + 5x^3 - 7$ is never decreasing and has just one stationary point. Deduce that the equation

$$x^5 + 5x^3 - 7 = 0$$

has just one real solution and, using trial calculations or otherwise, determine the value of this solution correct to one decimal place. Sketch the curve.

5. Sketch the curve

$$y = x^3(12x^2 - 15x - 40).$$

Answers

1. (i) $-\frac{1}{3} < x < 1$.
2. (i) $(\frac{1}{2}, \frac{3}{4})$ minimum.
   (ii) $(0, 3)$ maximum, $(1, 2)$ minimum.
   (iii) $(-3, -27)$ minimum, $(0, 0)$ horizontal inflexion.
   (iv) $(-2, 0)$ and $(2, 0)$ both minimum, $(0, 16)$ maximum.
   (v) $(-1, 12)$ maximum, $(0, 10)$ horizontal inflexion, $(1, 8)$ minimum.
   (vi) $(-2, 0)$ horizontal inflexion, $(-\frac{5}{7}, 4^4 \times 3^{10}/7^7)$ maximum, $(1, 0)$ minimum.
   (vii) $(-\frac{2}{3}, -2/3\sqrt{3})$ minimum.
   (viii) $(-2, 0)$ horizontal inflexion, $(1, \frac{27}{4})$ minimum.
4. 1.0.

6.3 THE ABSOLUTE MAXIMUM AND MINIMUM OF A FUNCTION

The terms **absolute maximum** and **absolute minimum** of a function are used to describe respectively the *greatest* and *least* values of the function in a given closed interval. When a function $f(x)$ is defined in a *closed* interval $x_1 \leqslant x \leqslant x_2$, the values of the function at a maximum (or minimum) statationary point within the interval may not necesssarily be the *absolute* maximum (or minimum) of the function in the interval $x_1 \leqslant x \leqslant x_2$. Additional consideration must be given to points where $f'(x)$ may not exist and to the end points $x = x_1$, $x = x_2$. See Fig. 6.4 where $f(x_1)$ is the absolute maximum of $f(x)$ in $x_1 \leqslant x \leqslant x_2$ and $f(x_3)$ is the absolute minimum ($f'(x_3)$ does not exist).

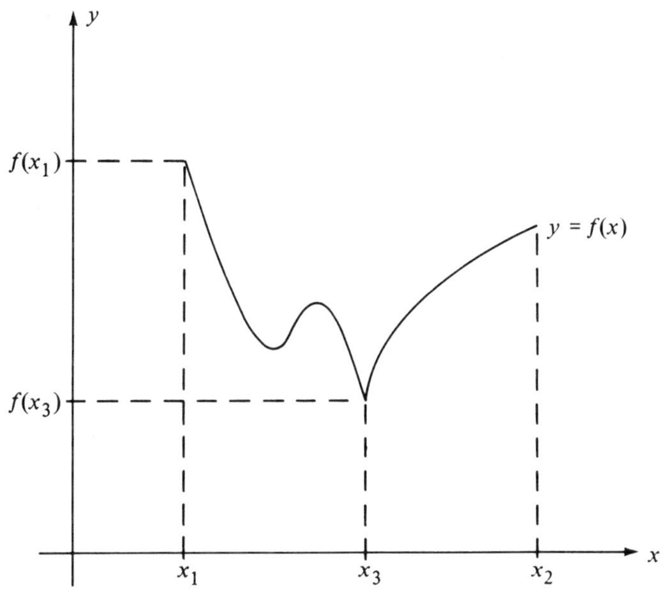

Fig. 6.4

Example 6.4

The function $y = |x(2x^2 + 9x + 12)|$ is defined in the closed interval $-1.5 \leqslant x \leqslant 1$. Determine the greatest and least value of y in this interval.

We note that the quadratic $2x^2 + 9x + 12$ is always positive (since it can be written $2\{(x + \frac{9}{4})^2 + \frac{15}{16}\}$). Therefore the function $x(2x^2 + 9x + 12)$ has the same sign as x. It follows that given function $y = |x(2x^2 + 9x + 12)|$ can be written in the form

$$y = \begin{cases} x(2x^2 + 9x + 12), & \text{when } x \geqslant 0, \\ -x(2x^2 + 9x + 12), & \text{when } x < 0. \end{cases}$$

Differentiating gives

$$\frac{\mathrm{d}y}{\mathrm{d}x} = \begin{cases} 6x^2 + 18x + 12, & \text{when } x > 0, \\ -(6x^2 + 18x^2 + 12), & \text{when } x < 0. \end{cases}$$

($\mathrm{d}y/\mathrm{d}x$ is *not defined* when $x = 0$). Stationary points occur when $\pm(6x^2 + 18x + 12) = 0$, i.e. when $x = -2, x = -1$. We disregard $x = -2$ since it is not in the given closed interval. As x increases through the value $x = -1$, $\mathrm{d}y/\mathrm{d}x = -6(x + 1)(x + 2)$ goes $+ \; 0 \; -$ and the point $(-1, 5)$ is a *maximum* stationary point. However, since we require the greatest and least values of y in the closed interval $-1.5 \leqslant x \leqslant 1$, we must also consider the end points of the interval and the point where $\mathrm{d}y/\mathrm{d}x$ is not defined. When $x = -1.5$, $y = |(-1.5)(3)| = 4.5$. When $x = 0$, ($\mathrm{d}y/\mathrm{d}x$ not defined), $y = 0$. When $x = 1$, $y = |(1)(23)| = 23$.

It follows that, in the closed interval, the *absolute maximum* value of y is 23 and the *absolute minimum* value is 0. Fig. 6.5 shows the graph of y versus x for $-1.5 \leqslant x \leqslant 1$.

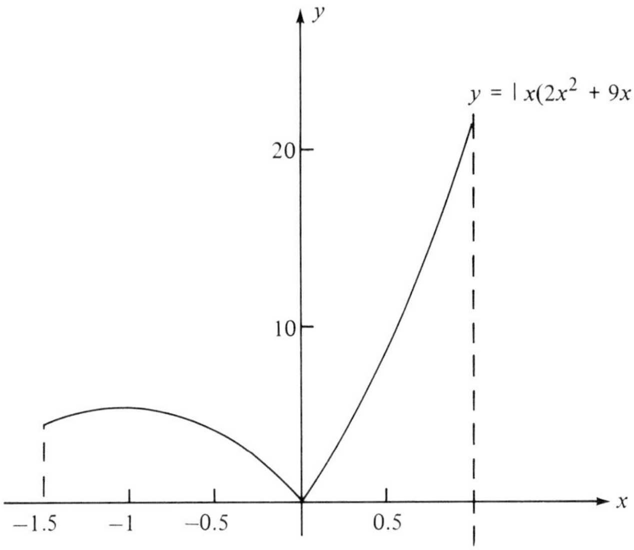

Fig. 6.5

Example 6.5

The perimeter of a rectangular field is fixed at 800 m, but a constraint exists which does not allow the shorter side to exceed 100 m. Calculate the greatest area for a field satisfying these conditions. What would the greatest area be if the constraint on the length of the shorter side is removed?

Denote the length of the shorter side by x and the length of the other side will be $400 - x$. The area A of the field is then given by

$$A = x(400 - x),$$

but x is constrained to the closed interval $0 \leqslant x \leqslant 100$. We require the absolute maximum value of A in this closed interval and proceed as in Example 6.4 by consideration of stationary points and interval end points (there are no points where $\mathrm{d}A/\mathrm{d}x$ is not defined).

$\mathrm{d}A/\mathrm{d}x = 400 - 2x = 0$ when $x = 200$, but $x = 200$ is *not* within the closed interval $0 \leqslant x \leqslant 100$. When $x = 0$, $A = 0$. When $x = 100$, $A = 30\,000$. Therefore the maximum value of A in the closed interval is $30\,000$ m^2.

When the constraint $x \leqslant 100$ is removed, the absolute maximum of A occurs at the single stationary point, where $x = 200$. Therefore the field is square with area $40\,000$ m^2.

Fig. 6.6 shows the graph of A versus x.

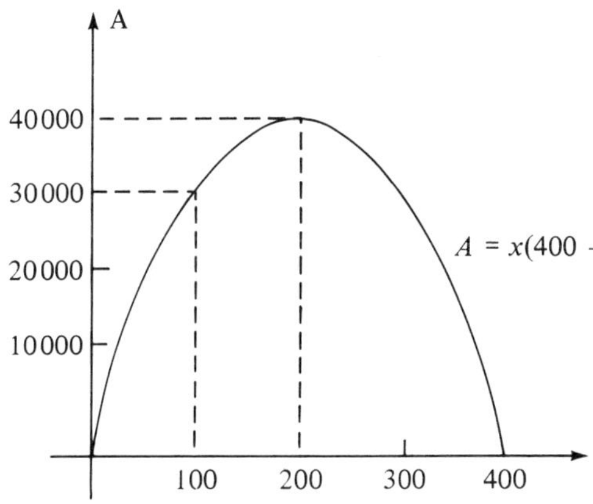

Fig. 6.6

Problems

1. Find the absolute maximum and the absolute minimum of the following functions in the given intervals.
   (i) $y = x(2x - 3)^2, \; -\frac{1}{2} \leqslant x \leqslant \frac{3}{2}$.
   (ii) $y = 4 - |x|, \; -5 \leqslant x \leqslant 5$.
   (iii) $y = |4 - x^2|, \; -3 \leqslant x \leqslant 3$.

2. Find the greatest and least values of the function $f(t) = (t-1)^2(t-2)$ in the interval $0.1 \leqslant t \leqslant 1.9$.

Answers

1. (i) 2, −8.
   (ii) 4, −1.
   (iii) 5, 0.

2. 0, −1.539.

6.4 APPLICATION TO OPTIMIZATION PROBLEMS

The theory set out in section 6.2 provides a powerful technique for solving problems which require the maximizing or minimizing of a certain quantity. In all such problems, we attempt to express the quantity as a function of *one* variable. The derivative with respect to this variable is equated to zero and solution of this equation leads to the evaluation of the required maximum or minimum. The first- or second-derivative test may be used to confirm the maximum or minimum, but in most practical problems this is unnecessary.

In many problems the independent variable is constrained to lie in a certain closed interval and it is the *absolute* maximum or minimum which is required. Such a problem has already been encountered in Example 6.5.

Example 6.6

120 cm^2 of sheet metal is to be used to construct an open-top tank with a square base. Find the dimensions of the tank such that its volume is maximized. What is the maximum volume?

Let S cm^2 be the surface area of the tank and V cm^3 be its volume. We are given that $S = 120$. Let the square base of the tank have sides of dimension x cm and let the height of the tank be y cm, as shown in Fig. 6.7.

Therefore,

$$V = x^2 y \tag{6.1}$$

and

$$S = x^2 + 4xy = 120. \tag{6.2}$$

V is the quantity to be maximized; therefore, we express V in terms of *one* variable. This can be done by expressing y in terms of x from equation (6.2) and substituting into equation (6.1), i.e.

$$y = \frac{120 - x^2}{4x}$$

and

$$\begin{aligned} V &= \frac{x^2(120 - x^2)}{4x} \\ &= \tfrac{1}{4}(120x - x^3). \end{aligned}$$

Practical considerations show that x must lie in the interval $0 \leqslant x \leqslant \sqrt{120}$ (otherwise, either x or V is negative!). Our problem, then, has been reduced to finding the greatest value of the function $V = \frac{1}{4}(120x - x^3)$ in the closed interval $0 \leqslant x \leqslant \sqrt{120}$.

Since V takes the value *zero* at both end points and $\mathrm{d}V/\mathrm{d}x$ exists for all values of x, it follows that V must be greatest at a maximum stationary point *within* the closed interval.

Now $\mathrm{d}V/\mathrm{d}x = \frac{1}{4}(120 - 3x^2) = 0$ when $x = \sqrt{40} = 2\sqrt{10}$, the *only* solution in $0 \leqslant x \leqslant \sqrt{120}$. Therefore, there is a maximum stationary point at $(2\sqrt{10}, 40\sqrt{10})$ on the V versus x graph. It follows that the dimensions of the tank are $x = 2\sqrt{10}$ cm, $y = \sqrt{10}$ cm, and the maximum volume is $40\sqrt{10}$ cm^3.

Example 6.7

At midnight, ship A is 50 nautical miles due north of ship B. A is steaming due east at 10 knots while B is steaming due north at 12 knots. At what time are the ships closest and what is the distance between them at this time?

Fig. 6.8 shows A_0 and B_0, the positions of the ships at midnight, and A and B, their positions t h later. $s = AB$ is the distance in nautical miles between the ships at time t h. Since the ships are travelling with constant velocity, we have $A_0A = 10t$ nautical miles and $B_0B = 12t$ nautical miles.

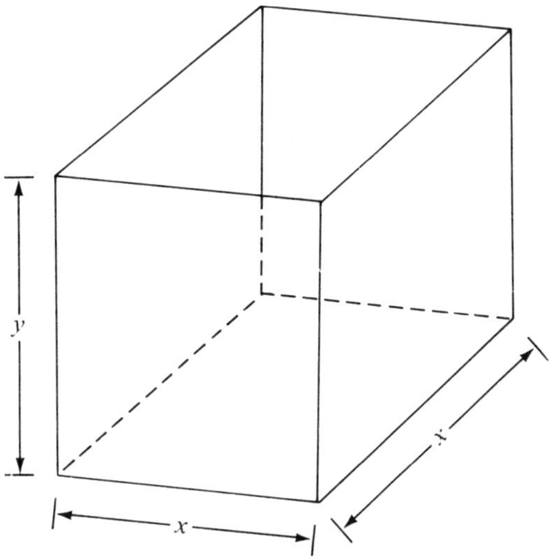

Fig. 6.7

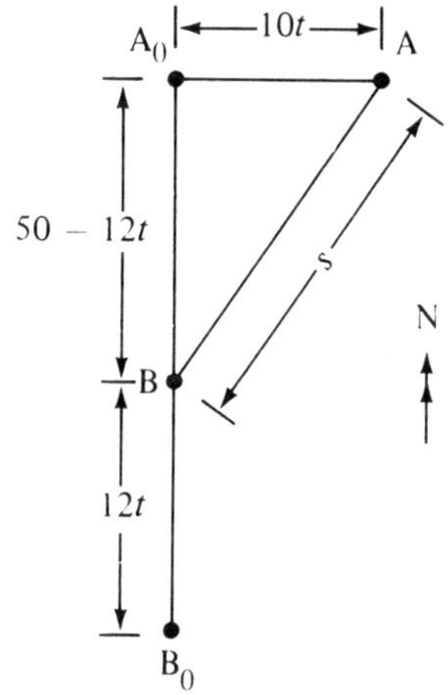

Fig. 6.8

s is the quantity to be minimized and therefore we attempt to express s in terms of the single variable t. From the triangle A_0AB in Fig. 6.8,

$$s^2 = (10t)^2 + (50 - 12t)^2$$

$$= 244t^2 - 1200t + 2500.$$

It is clear from practical considerations that there will be just one instant in $0 < t < \infty$ when the distance between the ships takes its least value. Further, since ds/dt is defined for all values of t, the closest distance must occur at a minimum stationary point on the s versus t curve. The normal procedure would be to solve the equation $ds/dt = 0$ to obtain the stationary point but, since the minimum value of s will occur at the same instant as the minimum value of s^2, it is easier to find the required time from the stationary point on the s^2 versus t curve, i.e. we solve $(\mathrm{d}/\mathrm{d}t)(s^2) = 0$ (which avoids the problem of differentiating the square root of a function):

$$\frac{\mathrm{d}}{\mathrm{d}t}(s^2) = 488t - 1200$$

$$= 0 \qquad \text{when } t = 2.459 \text{ h.}$$

Therefore the time when the ships are closest is approximately $27\frac{1}{2}$ minutes past 2 o'clock and the distance between the ships at this time is given by

$$s = \sqrt{(24.59)^2 + (50 - 12 \times 2.459)^2} \approx 32 \text{ nautical miles.}$$

Example 6.8

Find the maximum volume for a cylinder which can be inscribed in a right circular cone of height h and base radius r given that

(i) the axis of the cylinder is coincident with the axis of the cone and

(ii) the height of the cylinder is not less than $\frac{1}{2}h$.

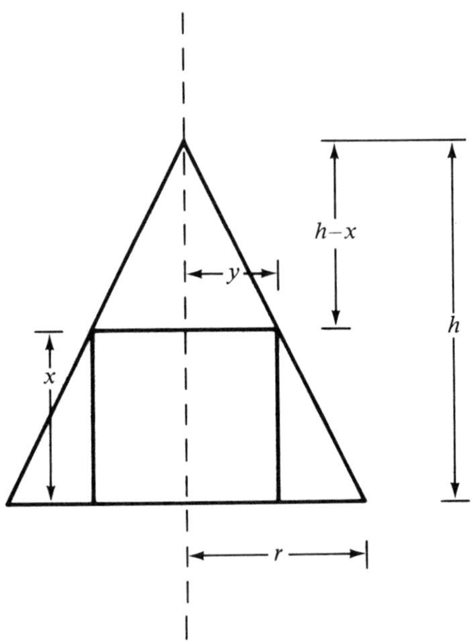

Fig. 6.9 Section through the axis of a cone and cylinder.

Fig. 6.9 shows a section through the common axis.

Let x be the height of the cylinder and y be its radius. x and y are variables but r and h are fixed. Let the volume of the cylinder be V. V has to be maximized and therefore must be expressed as a function of one variable.

$$V = \pi y^2 x. \tag{6.3}$$

By similar triangles,

$$\frac{r}{h} = \frac{y}{h - x}.$$

Therefore,

$$y = \frac{r}{h}(h - x). \tag{6.4}$$

Substitute equation (6.4) into equation (6.3) to obtain

$$V = \frac{\pi r^2}{h^2}(h - x)^2 x$$

$$= \frac{\pi r^2}{h^2}(h^2 x - 2hx^2 + x^3).$$

The problem has been reduced to one of maximizing $V(x)$ in the closed interval $\frac{1}{2}h \leqslant x \leqslant h$.

We note that $\mathrm{d}V/\mathrm{d}x$ is defined for all values of x and therefore, to obtain the *absolute* maximum of V, we require to consider stationary points on the V versus x curve and the values of V at the end points of the interval.

At stationary points,

$$\frac{\mathrm{d}V}{\mathrm{d}x} = \frac{\pi r^2}{h^2}(h^2 - 4hx + 3x^2) = 0,$$

i.e.

$$\frac{\pi r^2}{h^2}(h - x)(h - 3x) = 0,$$

and $x = h$, $x = \frac{1}{3}h$.

Therefore, there is no stationary point *within* the interval $\frac{1}{2}h \leqslant x \leqslant h$ and it is sufficient to consider the end points

$$V(\tfrac{1}{2}h) = \tfrac{1}{8}\pi r^2 h \quad \text{and} \quad V(h) = 0.$$

Therefore the required maximum volume is $\frac{1}{8}\pi r^2 h$.

Problems

1. Write down the maximum volume for the cylinder in Example 6.8 when the constraint (ii) is removed.

2. Design a cylindrical can to hold 1 litre such that the minimum material is used (1 litre = 1000 cm^3).

3. A beam of rectangular cross-section, depth d and breadth b, has stiffness proportional to bd^3. Find the dimensions of the stiffest such beam which can be cut from a length of wood of circular cross-section with a diameter D. Is this the heaviest beam?

4. The section of an open gutter made of thin material is to be rectangular and to have an area of 240 cm^2. If the base of the gutter is x cm, express the perimeter of the section in terms of x. Find the dimensions of the gutter if the perimeter is to be minimum.

5. A rectangular tank open at the top is to be made of sheet metal and to have a capacity of 30 m^3. If the width of the tank is 5 m, find the dimensions of the tank such that a minimum quantity of material is used. What is the area of sheet metal used for these dimensions?

6. Find the dimensions of the cone of maximum volume which has a slant height of 18 cm.

7. A rectangular field is to be fenced off along a road. The fence along the road costs £3 per metre while on the other sides it costs £2 per metre. Calculate the maximum area that can be fenced off for £400.

8. Fig. 6.10 shows points A and C on opposites sides of a straight river 1000 m wide. Point B is on the same side of the river as C such that CB = 500 m. A company requires to lay a pipeline so that oil may be pumped from A to B. The proposed pipeline consists of an underwater straight pipe from A to a point P on the opoosite bank and a land pipe from P to B. Determine the position of the point P such that the cost of the pipeline is minimized, given that underwater pipe laying is more expensive than land pipe by a factor of $\frac{5}{3}$.

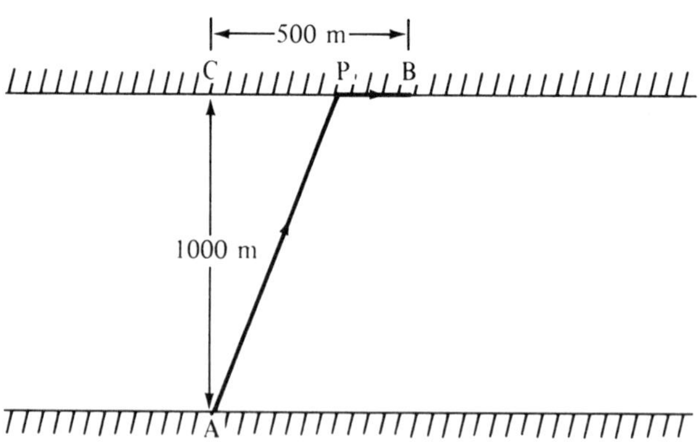

Fig. 6.10

9. A square sheet of metal of side a cm is to be used to make an open box by cutting a small square of metal from each corner and bending up the sides. What size should the cut-offs be for the box to have maximum volume?

10. Find the point on the curve $y = \sqrt{x}$ which is nearest to the point $(c, 0)$.

11. A manufacturer has enough factory space to produce 6000 toys of a certain type per week. The cost for producing n toys is $C = 100n + 20\,000$ p, and n toys can be sold per week at a price $P = 400 - 0.02n$ p per toy. Determine the value of n which maximizes the profit. If the company expanded its factory space such that 12 000 toys per week could be produced what value of n would maximize the profits? (Assume that the same functions apply for C and P.)

12. Figure 6.11 shows a cantilever beam of length l and weight W. It is built into a wall at one end and simply supported at the other. The deflection y of the beam at a distance x from the built-in end satisfies the equation

$$\frac{d^2y}{dx^2} = -\frac{W}{8lEI}(4x^2 - 5lx + l^2)$$

and the conditions $y = dy/dx = 0$ when $x = 0$, where E and I are constants depending on the type of beam. How far from the built-in end does the maximum deflection occur?

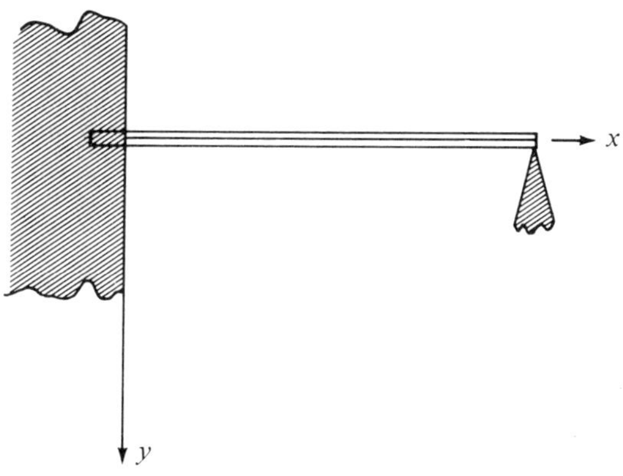

Fig. 6.11

Answers

1. $\dfrac{4\pi r^2 h}{27}$.

2. The diameter equals the height which equals $20\sqrt[3]{\dfrac{1}{2\pi}}$ cm.

3. $b = \dfrac{D}{2}$, $d = \dfrac{\sqrt{3}D}{2}$; no.
4. $\dfrac{480}{x} + x$; the base is 21.91 cm, and the height 10.95 cm approximately.
5. The base is 5 m $\times$ $2\sqrt{3}$ m, and the height $\sqrt{3}$ m; the area is 46.6 m^2 approximately.
6. The base radius is $6\sqrt{6}$ cm, and the height $6\sqrt{3}$ cm.
7. 2000 m^2.
8. P is coincident with B.
9. Squares $\dfrac{a}{6} \times \dfrac{a}{6}$.
10. $(c - \frac{1}{2}, \sqrt{c - \frac{1}{2}})$ when $c \geqslant \frac{1}{2}$; otherwise (0, 0).
11. 6000, 7500.
12. $0.5785l$.

6.5 POINTS OF INFLEXION

In Chapter 3 the concept of **concavity** was discussed. If $f''(x) > 0$ at a point on the curve $y = f(x)$, the curve lies above its tangent at the point and is said to be **concave up**. Similarly, if $f''(x) < 0$ at a point, the curve lies below its tangent and is said to be **concave down**.

A point on the curve $y = f(x)$ where the *concavity changes* is called a **point of inflexion**. The simple test for a point of inflexion is that $f''(x)$ *changes sign* as x increases through the point.

However, at a point of inflexion the curve should have a *unique* tangent line, which the curve crosses there. Thus we exclude points where the curve has a 'sharp corner' (i.e. points where $f'(x)$ has a sudden change in value) (Fig. 6.12). The unique tangent line at a point of inflexion may have a positive, negative or zero gradient (Fig. 6.13). If the gradient is zero, the point is *also* a stationary point which is called a *horizontal point of inflexion.*

To determine the positions of points of inflexion on the curve $y = f(x)$ it is necessary to find the points where $f''(x)$ *changes sign*. For 'smooth' curves (no sharp corners), this will happen when *either*

(i) $f''(x)$ passes through the value zero *or*
(ii) $f''(x)$ does not exist at the point.

For example,

(i) $y = x^3 + x$ has a point of inflexion at (0, 0) since $\mathrm{d}^2y/\mathrm{d}x^2 = 6x$ changes sign as x increases through $x = 0$. Note that $\mathrm{d}^2y/\mathrm{d}x^2 = 0$ at (0, 0).
(ii) $y = x^{5/3}$ has a point of inflexion at (0, 0) since $\mathrm{d}^2y/\mathrm{d}x^2 = \frac{10}{9}x^{-1/3}$ changes signs as x increases through $x = 0$. Note, however, that $\mathrm{d}^2y/\mathrm{d}x^2$ does not exist at (0, 0).

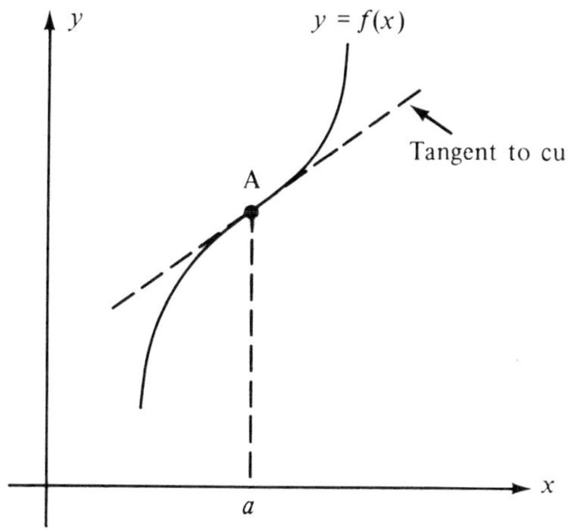

(a)

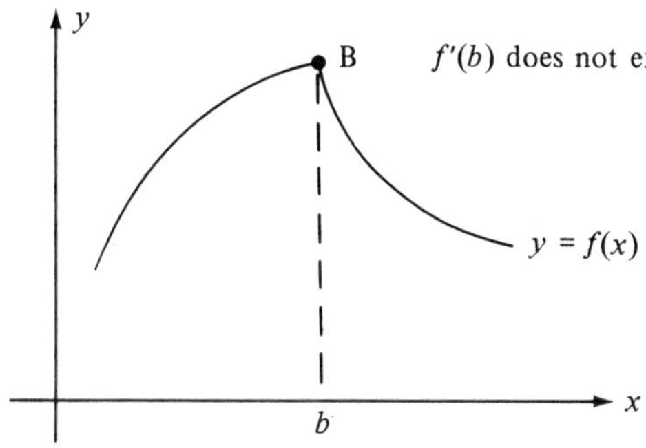

(b)

Fig. 6.12 (a) A is the point of inflexion, and the curve is smooth and crosses the tangent; (b) B is *not* a point of inflexion although concavity changes sign.

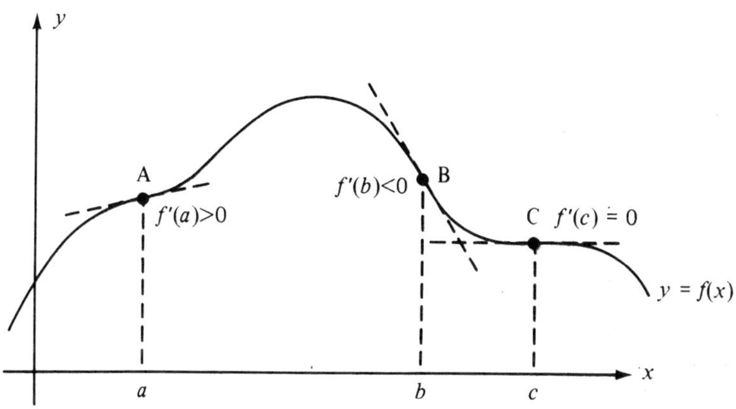

Fig. 6.13 A, B and C are points of inflexion. C is a horizontal poing of inflexion.

The graphs of $y = x^3 + x$ and $y = x^{5/3}$ are shown in Fig. 6.14.

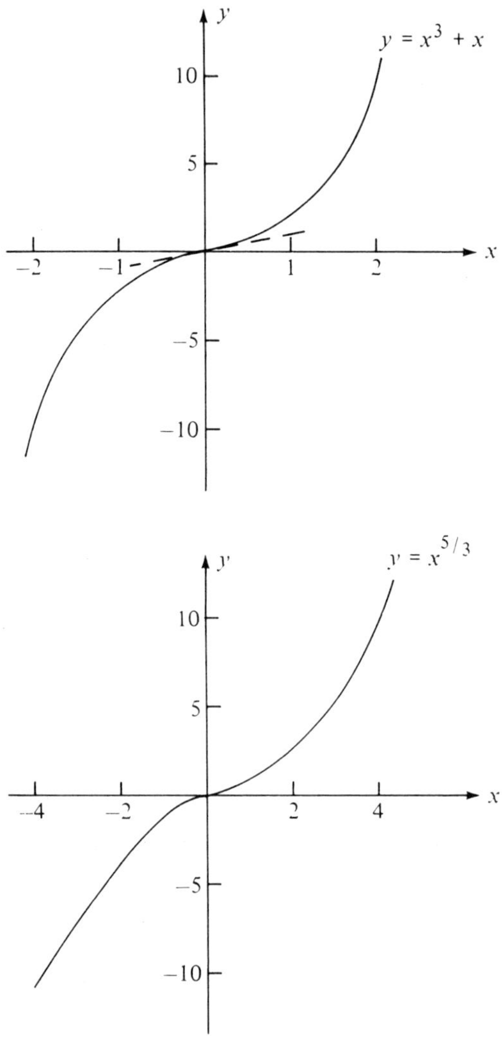

Fig. 6.14

Therefore, to locate a point of inflexion, we search for points where $d^2y/dx^2 = 0$ (or does not exist) and test each point for a change in sign of d^2y/dx^2. It is important to understand that a zero second derivative does *not* automatically imply a point of inflexion (Example 6.9 demonstrates that $d^2y/dx^2 = 0$ implies *nothing at all*).

Example 6.9

Show that $d^2y/dx^2 = 0$ at (0, 0) for all the following curves but that (0, 0) is a point of inflexion for only one of the curves.

(i) $y = x^4$.
(ii) $y = 5x + x^5$.
(iii) $y = |x^4 - x|$.
(iv) $y = x^4 + 4x$.

(i) $y = x^4$.

$$\frac{dy}{dx} = 4x^3.$$

$$\frac{d^2y}{dx^2} = 12x^2 = 0 \quad \text{at} \quad (0, 0).$$

d^2y/dx^2 does not change sign as x increases through $x = 0$ and $(0, 0)$ is *not* a point of inflexion. It is easy to confirm that $(0, 0)$ is a minimum stationary point.

(ii) $y = 5x + x^5$.

$$\frac{dy}{dx} = 5 + 5x^4.$$

$$\frac{d^2y}{dx^2} = 20x^3 = 0 \quad \text{at} \ (0, 0).$$

The sign of d^2y/dx^2 changes from $-$ to $+$ as x increases through $x = 0$ and $(0, 0)$ *is* a point of inflexion.

(iii) $y = |x^4 - x|$,
i.e.

$$y = \begin{cases} x^4 - x, & \text{when } x \leqslant 0 \text{ or } x \geqslant 1, \\ -x^4 + x, & \text{when } 0 < x < 1. \end{cases}$$

$$\frac{dy}{dx} = \begin{cases} 4x^3 - 1, & \text{when } x < 0 \text{ or } x > 1, \\ -4x^3 + 1, & \text{when } 0 < x < 1. \end{cases}$$

(Note that dy/dx does not exist when $x = 0$ since $(dy/dx)_{0+} \neq (dy/dx)_{0-}$; similarly, dy/dx does not exist when $x = 1$.)

$$\frac{d^2y}{dx^2} = \begin{cases} 12x^2, & \text{when } x \leqslant 0 \text{ or } x > 1, \\ -12x^2, & \text{when } 0 < x < 1. \end{cases}$$

Therefore, $d^2y/dx^2 = 0$ when $x = 0$ *and* changes sign as x increases through $x = 0$. However $(0, 0)$ is *not* a point of inflexion since the curve has a sharp corner at this point (dy/dx is not defined).

(iv) $y = x^4 + 4x$.

$$\frac{dy}{dx} = 4x^3 + 4.$$

$$\frac{d^2y}{dx^2} = 12x^2 = 0 \text{ at } (0, 0).$$

The function for d^2y/dx^2 is identical with that in (i) and (0, 0) is *not* a point of inflexion. However, in this case, (0, 0) is not a stationary point since the gradient has a value of 4 at the point.

The reader should carry out the curve-sketching procedure of section 0.3 for the curves in (i)–(iv) above. The corresponding sketches are shown in Figs 6.15(a)–6.15(d).

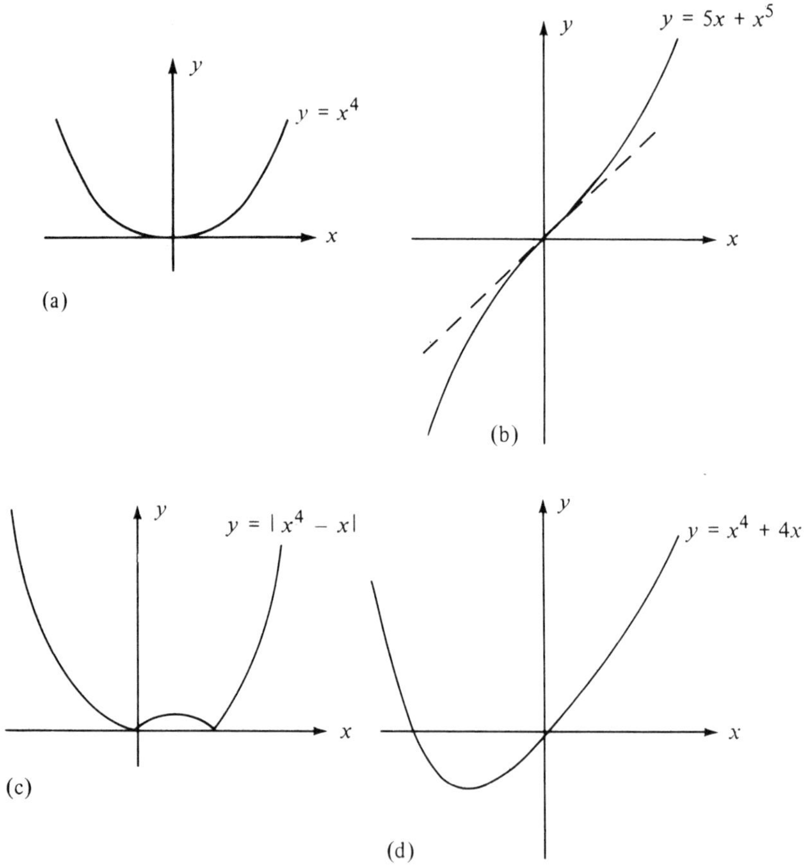

Fig. 6.15

Example 6.10

Show that the curve

$$y = 3x^5 - 10x^4 + 3x + 10$$

has just one point of inflexion.

$$\frac{dy}{dx} = 15x^4 - 40x^3 + 3.$$

$$\frac{d^2y}{dx^2} = 60x^3 - 120x^2$$

$$= 60x^2(x-2).$$

Therefore, $d^2y/dx^2 = 0$ when $x = 0$ and when $x = 2$.

As x increases through $x = 0$, d^2y/dx^2 does not change sign (it remains negative, i.e. the curve is concave down). Therefore, (0, 10) is *not* a point of inflexion.

As x increases through $x = 2$, d^2y/dx^2 changes from $-$ to $+$. Therefore, $(2, -48)$ is a point of inflexion and is the only one.

Example 6.11

Find the coordinates of the point of inflexion of the curve

$$y = (2x^2 - x^3)^{1/3}.$$

$$\frac{dy}{dx} = \tfrac{1}{3}(2x^2 - x^3)^{-2/3}(4x - 3x^2).$$

$$\frac{d^2y}{dx^2} = -\tfrac{2}{9}(2x^2 - x^3)^{-5/3}(4x - 3x^2)^2 + \tfrac{1}{3}(2x^2 - x^3)^{-2/3}(4 - 6x)$$

$$= \tfrac{1}{9}(2x^2 - x^3)^{-5/3}\{-2(4x - 3x^2)^2 + 3(2x^2 - x^3)(4 - 6x)\}$$

$$= \frac{-8x^2}{9(2x^2 - x^3)^{5/3}}$$

$$= \frac{-8}{9x^{4/3}(2-x)^{5/3}}.$$

d^2y/dx^2 is not defined at the points on the curve (0, 0) and (2, 0), and is zero nowhere.

As x increases through $x = 0$, d^2y/dx^2 remains negative. Therefore, (0, 0) is *not* a point of inflexion.

As x increases through $x = 2$, d^2y/dx^2 changes from $-$ to $+$. Therefore, (2, 0) is the only point of inflexion on the curve.

Example 6.12

Sketch the curve

$$y = \frac{4x}{1 + x^2},$$

showing clearly any stationary point and point of inflexion.

We follow the procedure of section 0.3, and additionaly search for points of inflexion.

(a) **Restrictions.** x may take any value, but y will lie within a closed interval. This can be determined algebraically but will be clearly defined when stationary points are considered.

(b) **Symmetry.** The curve is symmetrical about the origin. This follows since, if (a, b) is on the curve, then so is $(-a, -b)$.

(c) **Axis intersections.** (0, 0) only.

(d) **Stationary points and points of inflexion.**

$$\frac{dy}{dx} = \frac{4(1-x^2)}{(1+x^2)^2}.$$

$$\frac{d^2y}{dx^2} = \frac{8x(x^2-3)}{(1+x^2)^3}.$$

$$\frac{dy}{dx} = 0 \qquad \text{when } x = \pm 1.$$

The point (1, 2) is a maximum stationary point since $d^2y/dx^2 < 0$ (Theorem 6.2). By symmetry the point $(-1, -2)$ is a minimum stationary point. $d^2y/dx^2 = 0$ when $x = 0$, $x = \pm\sqrt{3}$. As x increases in turn through each one of these values, there is a change in sign of d^2y/dx^2. Therefore, (0, 0), $(\sqrt{3}, \sqrt{3})$, $(-\sqrt{3}, -\sqrt{3})$ are all points of inflexion. The gradients at these points are respectively 4, -0.5, -0.5.

(e) **Asymptotes.** $y = 0$; as $x \to \pm\infty$, $y \to 0$.

The sketch is shown in Fig. 6.16.

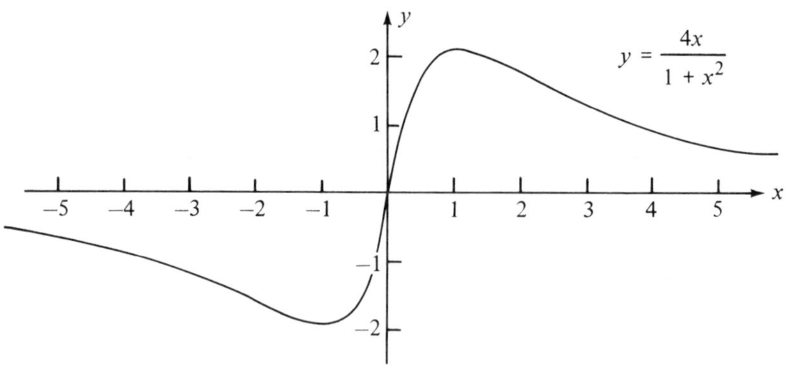

Fig. 6.16

Problems

1. (i) Show that $y = x + 36x^2 - 2x^3 - x^4$ has points of inflexion when $x = 2$ and $x = -3$.

(ii) Show that $y = x^3 - x^2 + x + 4$ has a point of inflexion when $x = \frac{1}{3}$.
(iii) Determine the points of inflexion of the curve

$$y = 5x^7 - 21x^5 + 35x^3 + 6x - 3.$$

(iv) Show that the curve $y = 5x^8 + 20x^7 + 21x^6 - 17$ has no point of inflexion.

2. Find the points of inflexion of the following curves.

(i) $$y = \frac{x^3}{1 + x^2}.$$

(ii) $y = (3 - x)^{5/3} + 3$.

3. Show that the curve

$$y = \left(\frac{x - 3}{x - 2}\right)^2$$

has a point of inflexion when $x = 3.5$. Sketch the curve.

4. Show that the curve

$$y = \frac{(x - 1)^2}{x + 1}$$

has no point of inflexion. Sketch the curve.

5. The beam described in problem 12 of section 6.4 deflects into a configuration given by solving for y in terms of x. The internal bending stress in the beam is approximately zero at a point corresponding to the point of inflexion of the configuration. How far from the built-in end does this point occur?

Answers

1. (iii) $(0, -3)$.
2. (i) $(0, 0)$, $(\sqrt{3}, 3\sqrt{3}/4)$, $(-\sqrt{3}, -3\sqrt{3}/4)$.
   (ii) $(3, 3)$.
5. $0.25l$.

Further Problems for Chapter 6

1. Determine the ranges of values of x for which the function $6 - 12x + 9x^2 - 2x^3$ is increasing.

2. Show that the curve $y = x - \sin 2x$ has a stationary point when $x = \pi/6$. What type is the stationary point?

3. Give an example to *disprove* the incorrect statement '$f''(x)$ must be negative at a maximum stationary point of the curve $y = f(x)$'.

4. Show that the point (0, 1) is a horizontal point of inflexion of the curve $y = 9\sqrt[3]{x^5} + 1$. Does $\mathrm{d}^2y/\mathrm{d}x^2 = 0$ at this point?

5. For a particular curve, $\mathrm{d}y/\mathrm{d}x = x(x-1)^2$. What type of stationary point occurs
   (i) at $x = 0$ and
   (ii) at $x = 1$?

6. Write down the greatest and least values of the function $x^2 - 2x + 5$ in the closed interval $0 \leqslant x \leqslant 3$.

7. What are the greatest and least values of the function $x^2 - 2x + 5$ in the interval $3 \leqslant x \leqslant 6$?

8. Determine the absolute maximum and absolute minimum of the function $y = |t^2 - 1|$ in the closed interval $-2 \leqslant t \leqslant 2$.

9. By drawing a number of rough sketches, convince yourself that the following theorem is true.

Theorem 6.3

The absolute maximum (minimum) of a *differentiable* function $f(x)$ in the closed interval $a \leqslant x \leqslant b$ must occur either at $x = a$, or at $x = b$, or at a maximum (minimum) stationary point of $f(x)$ in $a \leqslant x \leqslant b$.

10. The vertices of a rectangle lie on a circle of radius R. What is the maximum area for such a rectangle?

11. Given that $f(0) = 0, f'(0) = -1$ and $f''(0) = 2$, make a rough sketch of the curve $y = f(x)$ in the neighbourhood of the point (0, 0).

12. The curve $y = f(x)$ is such that

$$\frac{\mathrm{d}^2y}{\mathrm{d}x^2} = x^2(x+1)^3.$$

Which *one* of the following is true?
   (i) $y = f(x)$ has two points of inflexion.
   (ii) $y = f(x)$ has one point of inflexion, at $x = 0$.
   (iii) $y = f(x)$ has one point of inflexion, at $x = -1$.
   (iv) $y = f(x)$ has no point of inflexion.

13. Show that the curve $y = \sin x$ has a point of inflexion at (0, 0).

14. Evaluate d^2y/dx^2 at the point (0, 0) on the curve $y = \sin^4 x$. Is (0, 0) a point of inflexion?

Answers

1. $1 < x < 2$.
2. Minimum.
3. For example, $y = -x^4$.
4. No.
5. (i) Minimum.
   (ii) Horizontal inflexion.
6. 8, 4.
7. 29, 8.
8. 3, 0.
10. $2R^2$.
11. 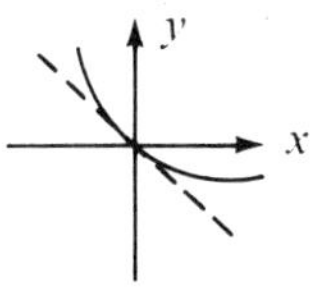

12. (iii) is true.
14. 0, no.

7

Applications of the Function of a Function Rule

7.1 INTRODUCTION

In Chapter 1 the function of a function rule, equation (1.8), was introduced as a powerful method for differentiating *explicit functions* by extending standard results (an explicit function of x is one which may be written in the form $y = f(x)$. If the equation of a curve is given as an explicit function, then the gradient of the curve at a particular point is obtained simply by evaluating the derivative of the function at the point. In sections 7.2 and 7.3, equation (1.8) is applied to determine the gradient of a curve when the equation of the curve is *not* in explicit function form.

A further application of equation (1.8) is the calculation of rates of change with respect to time, when two or more variables are related by some functional law. Problems of this type are presented in section 7.4.

7.2 DIFFERENTIATION OF IMPLICIT FUNCTIONS

Consider the equation $y^5 + 3y^2x - 12x^2 = 4$. It is not possible to rearrange the equation into an explicit form $y = f(x)$, but there is an *implied* functional relationship between x and y (i.e. a graph can be drawn by selecting a range of values for y and solving for x). The term **implicit function** is used to describe such a situation.

When x and y are related by an implicit function, we may obtain $\mathrm{d}y/\mathrm{d}x$ by differentiating the implicit function equation term by term with respect to x. It is necessary to use the function of a function rule in the special form

$$\frac{\mathrm{d}}{\mathrm{d}x}\{F(y)\} = \frac{\mathrm{d}}{\mathrm{d}y}\{F(y)\}\,\frac{\mathrm{d}y}{\mathrm{d}x},$$

i.e.

$$\frac{dF}{dx} = \frac{dF}{dy}\frac{dy}{dx}.$$

Returning to the example $y^5 + 3y^2x - 12x^2 = 4$, we differentiate the equation *term by term* with respect to x; this process is known as **implicit differentiation.**

Differentiation of the term y^5 gives

$$\frac{d}{dx}(y^5) = \frac{d}{dy}(y^5)\frac{dy}{dx}$$

$$= 5y^4\frac{dy}{dx}.$$

Differentiation of the term $3y^2x$ gives using equation (2.1), the product rule

$$\frac{d}{dx}(3y^2x) = (3y^2)(1) + \left\{\frac{d}{dx}(3y^2)\right\}(x)$$

$$= 3y^2 + \frac{d}{dy}(3y^2)\frac{dy}{dx}x$$

$$= 3y^2 + 6yx\frac{dy}{dx}.$$

Differentiation of the term $12x^2$ gives

$$\frac{d}{dx}(12x^2) = 24x.$$

Finally, we differentiate the right-hand side of the equation to obtain $(d/dx)(4) = 0$.

Therefore the implicit differentiation process, when applied to $y^5 + 3y^2x - 12x^2 = 4$, gives

$$5y^4\frac{dy}{dx} + 3y^2 + 6yx\frac{dy}{dx} - 24x = 0,$$

which can be arranged to obtain dy/dx in terms of x and y, i.e.

$$(5y^4 + 6yx)\frac{dy}{dx} = 24x - 3y^2$$

and

$$\frac{dy}{dx} = \frac{24x - 3y^2}{5y^4 + 6yx}.$$

It is important to understand that *all* differentiations involve this process of applying the operation of differentiation to *both sides* of an equation. Even when differentiating a simple explicit function such as $y = x^2$, the left-hand side y is differentiated with respect to x to obtain dy/dx and the right-hand side is differentiated also with respect to x to obtain $2x$. Thus, $dy/dx = 2x$.

Example 7.1
Determine the gradient of the curve $x^2 + xy + y^2 = 7$ at the point $(1, 2)$.

We differentiate the equation $x^2 + xy + y^2 = 7$ term by term with respect to x, noting that the product rule is required for the term xy. Therefore,

$$2x + \left\{x\frac{dy}{dx} + (y)(1)\right\} + 2y\frac{dy}{dx} = 0$$

and

$$\frac{dy}{dx} = -\frac{2x + y}{x + 2y}.$$

It follows that the gradient of the curve at the point $(1, 2)$ is $-\frac{4}{5}$.

Example 7.2
Obtain the equation of the tangent to the curve $4y = \sin(2x + 3y)$ at the point $(\pi/2, 0)$.

Differentiating the equation $4y = \sin(2x + 3y)$ with respect to x

$$4\frac{dy}{dx} = \cos(2x + 3y)\frac{d}{dx}(2x + 3y)$$

$$= \left(2 + 3\frac{dy}{dx}\right)\cos(2x + 3y).$$

Therefore

$$\{4 - 3\cos(2x + 3y)\}\frac{dy}{dx} = 2\cos(2x + 3y)$$

and at the point $(\pi/2, 0)$, $7(dy/dx) = -2$. Therefore the gradient of the tangent at $(\pi/2, 0)$ is $-\frac{2}{7}$ and the equation of the required tangent is

$$y - 0 = -\tfrac{2}{7}\left(x - \frac{\pi}{2}\right),$$

i.e. $2x + 7y - \pi = 0$.

Example 7.3
Use implicit differentiation to show that

$$\frac{d}{dx}(x^n) = nx^{n-1}$$

under the following conditions.

(i) When n is a positive rational number.

(ii) When n is a negative rational number.

(i) The rule for differentiating x^n when n is a *positive integer* was derived in Chapter 1, i.e.

$$\frac{d}{dx}(x^n) = nx^{n-1},$$

$$n = 1, 2, 3, \ldots \tag{1.5}$$

We use the equation (1.5) to prove the required result when n is a positive rational number. Let $n = p/q$ where p and q are positive integers. Let $y = x^n = x^{p/q}$, i.e. $y^q = x^p$. Applying the process of implicit differentiation to $y^q = x^p$, and using equation (1.5),

$$qy^{q-1}\frac{dy}{dx} = px^{p-1}.$$

i.e.

$$\begin{aligned}\frac{dy}{dx} &= \frac{p}{q}x^{p-1}y^{1-q}\\ &= \frac{p}{q}x^{p-1}(x^{p/q})^{1-q}\\ &= \frac{p}{q}x^{p-1}x^{p/q-p}\\ &= \frac{p}{q}x^{p/q-1}\\ &= nx^{n-1}\end{aligned}$$

since $n = p/q$. Therefore,

$$\frac{d}{dx}(x^n) = nx^{n-1}, \tag{7.1}$$

n a positive rational number.

(ii) n is now a *negative* rational number. Let $m = -n$ so that m is a positive rational number. Let $y = x^n = x^{-m}$, i.e. $x^m y = 1$. Differentiating the equation $x^m y = 1$ with respect to x and applying equation (7.1),

$$x^m \frac{dy}{dx} + mx^{m-1}y = 0$$

i.e.

$$\frac{dy}{dx} = -mx^{m-1}yx^{-m}$$

$$= -mx^{m-1}x^{-m}x^{-m}$$

$$= -mx^{-m-1}$$

$$= nx^{n-1}.$$

Therefore,

$$\frac{d}{dx}(x^n) = nx^{n-1},$$

n a negative rational number.

Example 7.4

Given that $x^2 + 2y^2 = 1$, use implicit differentiation to obtain dy/dx and d^2y/dx^2.

$$x^2 + 2y^2 = 1.$$

Therefore,

$$2x + 4y\frac{dy}{dx} = 0,$$

i.e.

$$\frac{dy}{dx} = -\frac{x}{2y}. \tag{7.2}$$

To obtain d^2y/dx^2, equation (7.2) is differentiated with respect to x, i.e.

$$\frac{d^2y}{dx^2} = -\frac{d}{dx}\left(\frac{x}{2y}\right)$$

Applying equation (2.2), the quotient rule,

$$-\frac{d}{dx}\left(\frac{x}{2y}\right) = -\left\{\frac{(2y)(1) - (x)(2)(dy/dx)}{4y^2}\right\}.$$

From equation (7.2),

$$-\left(\frac{2y \times 1 - x \times 2(dy/dx)}{4y^2}\right) = -\left\{\frac{2y - 2x(-x/2y)}{4y^2}\right\}$$

$$= -\left(\frac{2y^2 + x^2}{4y^3}\right)$$

$$= -\frac{1}{4y^3},$$

since $x^2 + 2y^2 = 1$.

Problems

1. For the semicircle

$$x^2 + y^2 = a^2, \; y \geqslant 0,$$

use implicit differentiation to show that

$$\frac{dy}{dx} = -\frac{x}{y},$$

$$\frac{d^2y}{dx^2} = -\frac{a^2}{y^3}.$$

Check by expressing y as an explicit function and differentiating twice.

2. Obtain dy/dx for the following implicit functions.
   (i) $y^5 + 3xy^3 + x^2y = 4$.
   (ii) $x^2 + 3x^2y^3 + y^4 = 71$.
   (iii) $x^{-1} + y^{-1} + 2xy = 6$.
   (iv) $\sin y = x^2 + 2x$.
   (v) $\tan\left(\frac{\pi}{3} + x - 2y\right) = 3y^2$.
   (vi) $(x^3 + 2xy)^4 = (1 - 2x)^3$.

3. Find the gradient of the curve

$$x^3 + 2x^2y - 3y^2 + 4x + y - 4 = 0$$

at the point $(2, -1)$.

4. Find the equations of the tangent and normal to the curve

$$x^2y^2 + x^3 = y^3 - 3$$

at the point $(1, 2)$.

5. Show that the equation of the tangent at the point (x_0, y_0) on the ellipse

$$\frac{x^2}{a^2} + \frac{y^2}{b^2} = 1$$

is

$$\frac{xx_0}{a^2} + \frac{yy_0}{b^2} = 1.$$

Answers

2. (i) $-\left(\dfrac{3y^3 + 2xy}{5y^4 + 9xy^2 + x^2}\right).$

(ii) $-\left(\dfrac{2x + 6xy^3}{9x^2y^2 + 4y^3}\right).$

(iii) $\dfrac{2y - x^{-2}}{y^{-2} - 2x}.$

(iv) $\dfrac{2x + 2}{\cos y}.$

(v) $\dfrac{\sec^2(\pi/3 + x - 2y)}{6y + 2\sec^2(\pi/3 + x - 2y)}.$

(vi) $-\left\{\dfrac{3(1 - 2x)^2 + 2(x^3 + 2xy)^3 (3x^2 + 2y)}{4x(x^3 + 2xy)^3}\right\}.$

3. $-\frac{8}{15}$.
4. The tangent is $8y - 11x - 5 = 0$, and the normal is $8x + 11y - 30 = 0$.

7.3 FUNCTIONS DEFINED BY PARAMETRIC EQUATIONS

Another way of expressing a functional relationship between x and y is to express both in terms of a third variable. The resulting *two* equations are known as the **parametric equations** for the function.

For example, the equations $x = t^2$ and $y = 2t$ are the parametric equations for the parabola $y^2 = 4x$ (eliminate t to show this). Every point on the curve $y^2 = 4x$ has a value of t associated with it and it is possible to identify any point on the curve simply by stating the *single* variable t. We may refer to 'the point $t = 3$' when considering the point (9, 6) and '$t = -1$' is an alternative name for the point $(1, -2)$. $\mathrm{d}y/\mathrm{d}x$ may be obtained from the parametric equations by applying a form of the function of a function rule:

$$\frac{\mathrm{d}y}{\mathrm{d}x} = \frac{\mathrm{d}y}{\mathrm{d}t}\,\frac{\mathrm{d}t}{\mathrm{d}x},$$

i.e.

$$\frac{\mathrm{d}y}{\mathrm{d}x} = \frac{\mathrm{d}y/\mathrm{d}t}{\mathrm{d}x/\mathrm{d}t}. \tag{7.3}$$

(Note that $\mathrm{d}t/\mathrm{d}x = (\mathrm{d}x/\mathrm{d}t)^{-1}$ which may be proved from the definition of a derivative, equation (1.2).)

To determine $\mathrm{d}^2y/\mathrm{d}x^2$, we simply apply equation (7.3) to the parametric equations defining $\mathrm{d}y/\mathrm{d}x$ and x.

$$\frac{\mathrm{d}^2y}{\mathrm{d}x^2} = \frac{(\mathrm{d}/\mathrm{d}t)\,(\mathrm{d}y/\mathrm{d}x)}{\mathrm{d}x/\mathrm{d}t}. \tag{7.4}$$

Note that equation (7.4) is precisely equation (7.3) with y replaced by $\mathrm{d}y/\mathrm{d}x$.

Returning to the example $x = t^2$, $y = 2t$, we obtain

$$\frac{\mathrm{d}y}{\mathrm{d}x} = \frac{2}{2t} = \frac{1}{t}$$

from equation (7.3) and

$$\frac{\mathrm{d}^2y}{\mathrm{d}x^2} = \frac{(\mathrm{d}/\mathrm{d}t)\,(1/t)}{2t} = -\frac{1}{2t^3}$$

from equation (7.4).

In this simple example it is possible to eliminate t and to express y in terms of x, but frequently this is not possible.

A common application of parametric equations is in dynamics in two dimensions. The position of a moving body may be defined by expressing its coordinates x and y in terms of time t. The components of the velocity vector along the x and y axes are respectively $\mathrm{d}x/\mathrm{d}t$ and $\mathrm{d}y/\mathrm{d}t$, and $\mathrm{d}y/\mathrm{d}x$ must be calculated to obtain the direction of motion at time t (see Example 7.6).

***Example** 7.5*

Determine $\mathrm{d}y/\mathrm{d}x$ and $\mathrm{d}^2y/\mathrm{d}x^2$ for the function defined by the parametric equations

$$x = 3\cos^3\theta,$$

$$y = 3\sin^3\theta.$$

$$\frac{\mathrm{d}x}{\mathrm{d}\theta} = 9\cos^2\theta\,(-\sin\theta).$$

$$\frac{\mathrm{d}y}{\mathrm{d}\theta} = 9\sin^2\theta\cos\theta.$$

$$\frac{\mathrm{d}y}{\mathrm{d}x} = \frac{\mathrm{d}y/\mathrm{d}\theta}{\mathrm{d}x/\mathrm{d}\theta}$$

$$= \frac{9\sin^2\theta\cos\theta}{-9\cos^2\theta\sin\theta}$$

$$= -\frac{\sin\theta}{\cos\theta}$$

$$= -\tan\theta.$$

$$\frac{d^2y}{dx^2} = \frac{(d/d\theta)(dy/dx)}{dx/d\theta}$$

$$= \frac{-\sec^2\theta}{-9\cos^2\theta\sin\theta}$$

$$= \tfrac{1}{9}\sec^4\theta\,\mathrm{cosec}\,\theta.$$

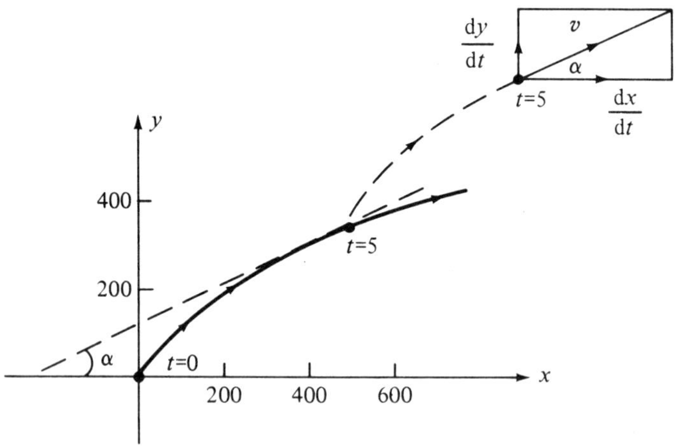

Fig. 7.1

Example 7.6
A projectile is launched from a point 0 with velocity $100\sqrt{2}$ m/s at an angle 45° above the horizontal. Its position t s later is given by the coordinates $x = 100t, y = 100t - 5t^2$ (Fig. 7.1). Determine the magnitude of the velocity vector and the direction of motion at the instant $t = 5$.

The components of the velocity vector are dx/dt and dy/dt and the direction of motion is along the tangent to the trajectory, i.e. at angle α above the horizontal where $\tan\alpha = dy/dx$ (see Fig. 7.1). The magnitude of the velocity vector at any instant is given by the value of

$$v = \sqrt{\left(\frac{dx}{dt}\right)^2 + \left(\frac{dy}{dt}\right)^2}.$$

$$\frac{dx}{dt} = 100$$

and

$$\frac{dy}{dt} = 100 - 10t.$$

Therefore,

$$\frac{dy}{dx} = \frac{100 - 10t}{100}.$$

When $t = 5$, we have

$$v = \sqrt{100^2 + 50^2}$$
$$\approx 111.8 \text{ m/s}.$$
$$\tan \alpha = \tfrac{50}{100} = 0.5$$

Therefore,

$$\alpha \approx 26.6°.$$

Problems

1. Find dy/dx for the following functions defined by parametric equations.
   (i) $x = 3t + 1$, $y = t^2$.

   (ii) $x = \dfrac{t}{1+t}$, $y = \dfrac{t^2}{1+t}$.

   (iii) $x = 1 - \cos 2\theta$, $y = 2\theta - \sin 2\theta$.

2. Find d^2y/dx^2 for the functions defined in problem 1.

3. Determine the values of dy/dx and d^2y/dx^2 when $x = \frac{1}{2}$ for the function $x = \sin t$, $y = t \sin t$, $0 \leqslant t \leqslant \pi/2$.

4. Find the equation of the tangent to the curve

$$x = \frac{t}{\sin t}, \quad y = \cos t$$

   at the point where $t = \pi/2$.

5. A particle moves in a plane such that its coordinates with respect to perpendicular axes Oxy are given by

$$x = a \cos \omega t, \quad y = a \sin \omega t$$

where a and ω are constants and t is time. Show that the magnitude of the particle's velocity is constant and that the direction of motion makes an angle $\pi/2 + \omega t$ with the axis Ox at time t. Describe the curve traced out by the moving particle.

Answers

1. (i) $\dfrac{2t}{3}$.

   (ii) $2t + t^2$.

   (iii) $\tan \theta$.

2. (i) $\frac{2}{9}$.

   (ii) $2(1 + t)^3$.

   (iii) $\dfrac{1}{4 \sin \theta \cos^3 \theta}$.

3. $\dfrac{\sqrt{3}}{3} + \dfrac{\pi}{6}, \; \dfrac{14\sqrt{3}}{9}$.

4. $y = -x + \dfrac{\pi}{2}$.

5. Particle describes a circle of centre (0, 0) and radius a.

7.4 RELATED RATES

In many applications, two variables are related by some functional law (e.g. $y = f(x)$) and both of the variables are dependent on a third variable, usually time t. The rate of change in y with respect to t may be related to the rate of change in x with respect to t by applying the function of a function rule, i.e.

$$\frac{\mathrm{d}y}{\mathrm{d}t} = \frac{\mathrm{d}y}{\mathrm{d}x}\frac{\mathrm{d}x}{\mathrm{d}t}.$$

In a typical example *one* of $\mathrm{d}y/\mathrm{d}t$ and $\mathrm{d}x/\mathrm{d}t$ will be given and the other will have to be determined at some instant (when x and y will be known). It will be necessary to establish a relationship between x and y and then to differentiate the equation with respect to t. This will provide a relationship between the rates of change in x and y, and the final step will be to insert the particular numbers corresponding to the given instant. In all problems of this type we follow the sequence distinguish between given rate and required rate

↓

establish a relationship between variables

↓

differentiate equation with respect to t

↓

insert particular values.

***Example** 7.7*

A ladder 13 m long is resting against a wall. If the foot of the ladder slips away from the base of the wall at a rate of 0.5 m/s, how fast is the top sliding down the wall when the foot is 5 m from the base of the wall?

Fig. 7.2 shows the position of the ladder at time t; x is the distance of the ladder from the wall and y is the height of the top of the ladder. Note that we do *not* put $x = 5$ on the diagram since x is a variable.

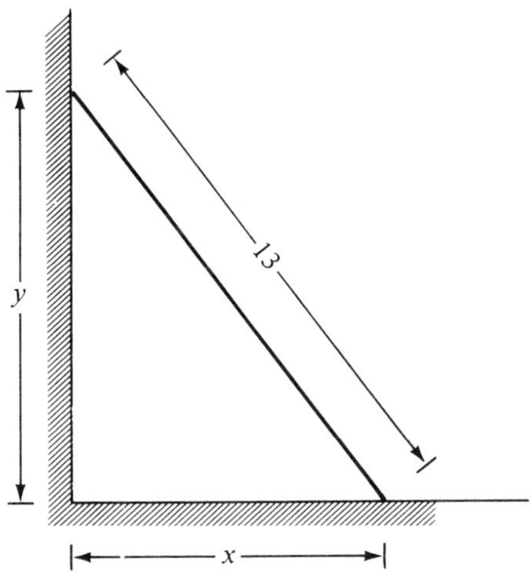

Fig. 7.2

The rate of change in x is *given* for all t, i.e. $\mathrm{d}x/\mathrm{d}t = 0.5$.

The rate $\mathrm{d}y/\mathrm{d}t$ of change in y is *required*. Therefore, we try to find a functional relationship between y and x which we can differentiate with respect to t using the function of a function rule.

From the right-angled triangle in Fig. 7.2,

$$x^2 + y^2 = 169. \tag{7.5}$$

Therefore $y = \sqrt{169 - x^2}$ and, differentiating with respect to t,

$$\frac{dy}{dt} = \frac{dy}{dx}\frac{dx}{dt}$$

$$= \frac{-x}{\sqrt{169-x^2}}\frac{dx}{dt}.$$

Finally, we insert the numbers which apply at the instant when $x = 5$, i.e.

$$\frac{dy}{dt} = -\frac{5}{\sqrt{144}} \times 0.5 = -\tfrac{5}{24}.$$

Therefore the top of the ladder is sliding *down* the wall at a rate $\frac{5}{24}$ m/s.

Note that it is possible to avoid the differentiation of the square root function by not expressing equation (7.5) in explicit form. We simply apply the implicit differentiation technique to equation (7.5) as follows:

$$2x\frac{dx}{dt} + 2y\frac{dy}{dt} = 0.$$

The numbers which apply when $x = 5$ are $y = \sqrt{169-25} = 12$ and $dx/dt = 0.5$. Therefore

$$10 \times 0.5 + 24\frac{dy}{dt} = 0$$

and

$$\frac{dy}{dt} = -\tfrac{5}{24}.$$

Example 7.8

Figure 7.3 shows a conical fuel tank which is 2.5 m deep and has a top diameter of 2 m. Fuel is withdrawn from the tank at a rate of 0.25 m^3/min. At what rate is the level of fuel falling at the instant when the depth of fuel is 1.5 m?

Note that Fig. 7.3 does not show numerical values for the depth h of fuel and the surface radius r, since these quantities are *variables*.

Let V be the volume of fuel in tank (V is also a *variable*). The rate of change in V is *given* for all time; $dV/dt = -0.25$ (negative because V is decreasing).

The rate dh/dt of change in h is *required*. Therefore, we must find a relationship between V and h which we can differentiate with respect to t.

$V = \frac{1}{3}\pi r^2 h$ is the formula for the volume of a cone. r may be eliminated since, by similar triangles,

$$\frac{r}{1} = \frac{h}{2.5},$$

i.e.

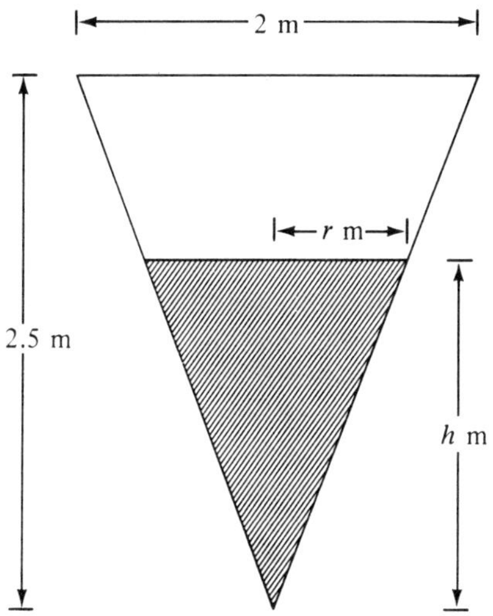

Fig. 7.3

$$r = \frac{2h}{5}$$

and

$$V = \tfrac{1}{3}\pi \left(\frac{2h}{5}\right)^2 h$$

$$= \frac{4\pi h^3}{75}.$$

Differentiating with respect to t,

$$\frac{\mathrm{d}V}{\mathrm{d}t} = \frac{\mathrm{d}V}{\mathrm{d}h}\,\frac{\mathrm{d}h}{\mathrm{d}t}$$

$$= \frac{4\pi h^2}{25}\,\frac{\mathrm{d}h}{\mathrm{d}t}.$$

$\mathrm{d}V/\mathrm{d}t = -0.25$ (given) and, at the instant when $h = 1.5$,

$$-0.25 = \frac{4\pi \times (1.5)^2}{25}\,\frac{\mathrm{d}h}{\mathrm{d}t}.$$

Therefore,

$$\frac{\mathrm{d}h}{\mathrm{d}t} = -\frac{6.25}{9\pi}$$

$$\approx -0.22 \text{ m/min},$$

i.e. the level of fuel is *falling* at approximately 0.22 m/min at the instant when the depth of fuel is 1.5 m.

Problems

1. Each side of an equilateral triangle is increasing at a rate of 2 cm/s. Find the rate at which its area is increasing when the side is of length $2\sqrt{3}$ cm.

2. A circular metal plate expands while being heated. The radius of the plate is increasing at a rate of 0.0001 mm/s. Find the rate at which the area of the plate is increasing when the radius is 20 cm.

3. Water flows into a cylindrical tank at the rate of 10 m^3/min. If the radius of the tank is 3 m, how fast is the water level rising?

4. Sand falls into a conical pile at a rate of 0.4 m^3/min. The diameter of the base of the pile is always equal to the height. How fast is the height of the pile rising at the instant when it is 2 m high?

5. A boat is pulled into dock by means of a rope with one end attached to the bow of the boat and the other end passing through a ring attached to the dock at a point 9 m higher than the bow. If the rope is pulled in at a rate of 1 m/s, how fast is the boat approaching the dock when it is 12 m from the dock.

6. A particle moves around the circle $x^2 + y^2 = 100$ such that its y coordinate is increasing at a rate of 8 m/s. What is the rate of change in the x coordinate when the x coordinate is 8 m?

7. Two railway tracks intersect at 60°. A train on one track is 10 miles from the junction moving towards it at 30 miles/h. Another train is on the other track, also 10 miles from the junction, but moving away from it at 20 miles/h. Calculate the rate of change in the distance between the trains.

8. Fig. 7.4 shows a searchlight situated at a distance 100 m from a straight wall. The searchlight is rotating at a constant rate of 1 rev/min and the beam of light strikes the wall at point P. Calculate the speed at which P travels along the wall under the following conditions.
   (i) When $\theta = 0$.
   (ii) When $x = 75$ m.

Answers

1. 6 cm^2/s.
2. 0.04π mm^2/s.
3. $\dfrac{10}{9\pi}$ m/min.

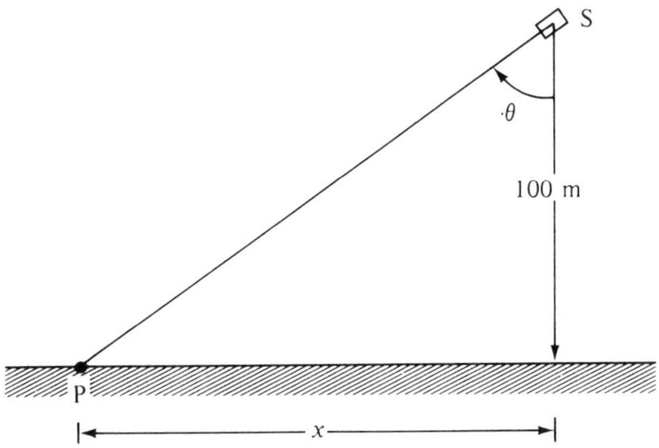

Fig. 7.4

4. $\dfrac{2}{5\pi}$ m/min.

5. 1.25 m/s.
6. −6 m/s.
7. −5 miles/h.
8. (i) 200π m/min.

   (ii) $\dfrac{625\pi}{2}$ m/min.

Further Problems for Chapter 7

1. Determine the gradient of the curve $x^2 + y^2 = 5$ at the point (1, 2).

2. A curve is described by the parametric equations $x = t + 2$, $y = t^2$. Calculate the gradient of the normal to the curve at the point $t = 2$.

3. Given that $\sin y = x + y$ show that

$$\frac{dy}{dx} = \frac{1}{\cos y - 1}.$$

4. Show that the tangent to the curve $x = 1 + \sin\theta$, $y = \theta + \cos\theta$ at the point (1, 1) is inclined at 45° to the x axis.

5. Given that $y = 10t - 5t^2$ express dy/dx in terms of t and dx/dt.

6. Given that $y^3 = 3x + 3y$ obtain an expression for d^2y/dx^2 in terms of x and y.

7. For the function $x = 1 + \sin\theta$, $y = \theta + \cos\theta$ evaluate d^2y/dx^2 at the point (1, 1).

8. A particle moves along the curve $y = x - 0.05x^2$ such that $\mathrm{d}x/\mathrm{d}t = 10$ m s^{-1}. Find the value of $\mathrm{d}y/\mathrm{d}t$ when $x = 5$ m.

9. A spherical balloon is inflated at a constant rate of 0.08 m^3/min. Calculate the rate of increase in the radius of the balloon when the radius is 0.1 m.

Answers

1. -0.5.
2. -0.25.
5. $10(1-t)\left(\dfrac{\mathrm{d}x}{\mathrm{d}t}\right)^{-1}$.
6. $-\dfrac{2y}{(y^2-1)^3}$.
7. -1.
8. 5 m/s.
9. $\dfrac{2}{\pi}$ m/min.

8

The Exponential, Logarithmic and Hyperbolic Functions

8.1 INTRODUCTION

In section 0.4 the number e was defined by the infinite series

$$e = 1 + \frac{1}{2} + \frac{1}{1 \times 2} + \frac{1}{1 \times 2 \times 3} + \frac{1}{1 \times 2 \times 3 \times 4} + \ldots$$

or, using the summation notation,

$$e = \sum_{n=0}^{\infty} \frac{1}{n!},$$

where $n!$, the *factorial function*, is defined for $n = 0, 1, 2, 3, 4, \ldots$ as follows:

$$0! = 1,$$

$$n! = n(n-1)!, \qquad n = 1, 2, 3, 4, \ldots,$$

i.e.

$$1! = 1,\ 2! = 2 \times 1,\ 3! = 3 \times 2 \times 1, \text{etc.}$$

The number e is a non-terminating and non-recurring decimal which correct to five decimal places is 2.718 28.

The function $\log_e x$, the logarithm to the base e of x, is called the *natural logarithm* of x and is written $\ln x$. The function $\ln x$ is defined for $x > 0$ and referred to as the *logarithmic function to the base* e or simply the **logarithmic function.** We recall from section 0.4 that $x = \log_a y$ and $y = a^x$ are different ways of writing the same relationship between x and y. It follows that $y = e^x$ is equivalent to $x = \ln y$, i.e. e^x is the *inverse natural logarithm* of x. The function e^x, sometimes written $\exp x$, is called the *exponential function with exponent* e or simply the **exponential function**. Its importance in applied mathematics is related to the property

$$\frac{d}{dx}(e^x) = e^x,$$

i.e. the operation of differentiation leaves the exponential function unchanged. This standard result will be derived in section 8.2 together with the derivative of the logarithmic function. Graphs of $y = \ln x$ and $y = e^x$ are shown in Fig. 8.1.

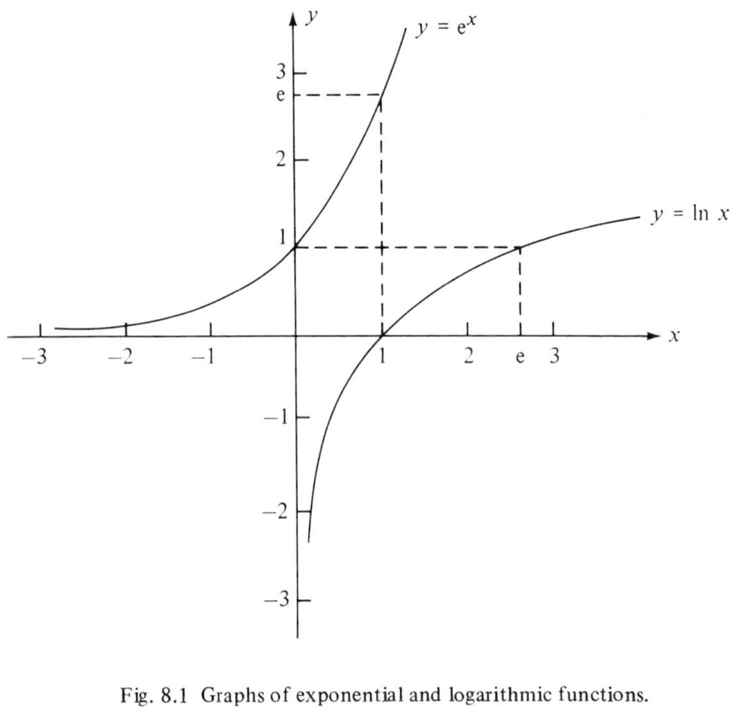

Fig. 8.1 Graphs of exponential and logarithmic functions.

In section 8.4, *hyperbolic functions* are defined in terms of exponential functions. The similarities between hyperbolic functions and trigonometric functions are emphasized.

8.2 DIFFERENTIATION OF e^x AND $\ln x$

The derivative of the logarithmic function $\ln x$, $x > 0$ is obtained by applying the definition of a derivative, equation (1.2), and using the laws of logarithms set out in section 0.4. These laws are repeated here for *natural logarithms*.

$$\ln x + \ln y = \ln(xy), \tag{8.1}$$

$$\ln x - \ln y = \ln\left(\frac{x}{y}\right), \tag{8.2}$$

$$n \ln x = \ln(x^n). \tag{8.3}$$

$$\frac{d}{dx}\{\ln x\} = \lim_{h\to 0}\left\{\frac{\ln(x+h)-\ln x}{h}\right\}$$

From equation (8.2),

$$\lim_{h\to 0}\left\{\frac{\ln(x+h)-\ln x}{h}\right\} = \lim_{h\to 0}\left\{\frac{\ln(1+h/x)}{h}\right\}$$

$$= \lim_{h\to 0}\left\{\frac{1}{x}\,\frac{x}{h}\ln\left(1+\frac{h}{x}\right)\right\}.$$

If $n = x/h$ is substituted, this equals

$$\lim_{n\to\infty}\left\{\frac{1}{x}n\ln\left(1+\frac{1}{n}\right)\right\}$$

From equation (8.3),

$$\lim_{n\to\infty}\left\{\frac{1}{x}n\ln\left(1+\frac{1}{n}\right)\right\} = \lim_{n\to\infty}\left\{\frac{1}{x}\ln\left(1+\frac{1}{n}\right)^n\right\}$$

$$= \frac{1}{x}\ln\left[\lim_{n\to\infty}\left\{\left(1+\frac{1}{n}\right)^n\right\}\right].$$

The number

$$\lim_{n\to\infty}\left\{\left(1+\frac{1}{n}\right)^n\right\}$$

may be evaluated using the binomial series given in section 13.3:

$$\left(1+\frac{1}{n}\right)^n = 1 + n\frac{1}{n} + \frac{n(n-1)}{2!}\left(\frac{1}{n}\right)^2 + \frac{n(n-1)(n-2)}{3!}\left(\frac{1}{n}\right)^3 + \ldots$$

$$= 1 + 1 + \frac{1}{2!}\left(1-\frac{1}{n}\right) + \frac{1}{3!}\left(1-\frac{1}{n}\right)\left(1-\frac{2}{n}\right) + \ldots.$$

Therefore,

$$\lim_{n\to\infty}\left\{\left(1+\frac{1}{n}\right)^n\right\} = 1 + 1 + \frac{1}{2!} + \frac{1}{3!} + \ldots$$

$$= e$$

and

$$\frac{d}{dx}(\ln x) = \frac{1}{x}\ln e.$$

Now $\ln e = 1$ since $e = e^1$ and we have the important result

$$\frac{d}{dx}(\ln x) = \frac{1}{x}, \qquad x > 0. \tag{8.4}$$

The derivative of the exponential function $y = e^x$ follows immediately from equation (8.4) using the inverse relationship $x = \ln y$, and differentiating with respect to y, i.e.

$$\frac{dx}{dy} = \frac{1}{y}$$
$$= \frac{1}{e^x}.$$

Therefore,

$$\frac{dy}{dx} = e^x$$

or

$$\frac{d}{dx}(e^x) = e^x. \tag{8.5}$$

The extended rules of differentiation for logarithmic and exponential functions follow by applying equation (1.8), the function of a function rule.

$$\frac{d}{dx}(\ln u) = \frac{1}{u}\,\frac{du}{dx}, \tag{8.6a}$$

$$\frac{d}{dx}(e^u) = e^u\,\frac{du}{dx}, \tag{8.7a}$$

or, alternatively, if u is the function $u = g(x)$.

$$\frac{d}{dx}\{\ln[g(x)]\} = \frac{1}{g(x)}g'(x), \tag{8.6b}$$

$$\frac{d}{dx}(e^{g(x)}) = e^{g(x)}\,g'(x). \tag{8.7b}$$

When using equation (8.6), consideration should be given before differentiation to the simplification of $\ln u$ by application of the laws of logarithms, equations (8.1)–(8.3). For example, if it is necessary to obtain $(d/dx)\{\ln(x\sqrt{1-x^2})\}$, then *before* applying equation (8.6b) the following simplification is made: from equation (8.1),

$$\ln(x\sqrt{1-x^2}) = \ln x + \ln(\sqrt{1-x^2}).$$

From equation (8.3),

$$\ln x + \ln(\sqrt{1-x^2}) = \ln x + \tfrac{1}{2}\ln(1-x^2).$$

Now the differentiation is completed using equation (8.6b):

$$\begin{aligned}\frac{\mathrm{d}}{\mathrm{d}x}\{\ln(x\sqrt{1-x^2})\} &= \frac{\mathrm{d}}{\mathrm{d}x}(\ln x) + \frac{\mathrm{d}}{\mathrm{d}x}\{\tfrac{1}{2}\ln(1-x^2)\}\\ &= \frac{1}{x} + \frac{1}{2}\frac{(-2x)}{(1-x^2)}\\ &= \frac{1}{x} - \frac{x}{1-x^2}.\end{aligned}$$

The above simplification avoided the problem of differentiating the *product* $x\sqrt{1-x^2}$ which would have been necessary had equation (8.6b) been applied directly.

The simplification afforded by taking logarithms may also be applied in situations requiring the differentiation of functions which involve complicated products, quotients and powers.

For example, to differentiate the function

$$\frac{x}{\sqrt{1-x^2}}$$

let

$$y = \frac{x}{\sqrt{1-x^2}}$$

and, from equation (8.2),

$$\ln y = \ln x - \ln(\sqrt{1-x^2})$$

From equation (8.3),

$$\ln x - \ln(\sqrt{1-x^2}) = \ln x - \tfrac{1}{2}\ln(1-x^2).$$

The equation

$$\ln y = \ln x - \tfrac{1}{2}\ln(1-x^2)$$

is now differentiated with respect to x using the implicit differentiation technique, i.e.

$$\frac{1}{y}\frac{\mathrm{d}y}{\mathrm{d}x} = \frac{1}{x} + \frac{x}{1-x^2}$$

and

$$\frac{\mathrm{d}y}{\mathrm{d}x} = y\left(\frac{1}{x} + \frac{x}{1-x^2}\right).$$

This result may be expressed entirely in terms of x by substituting for y, but frequently this is unnecessary. The technique of taking logarithms of a function in order to simplify its differentiation is known as **logarithmic differentiation**.

The standard integration rules corresponding to equations (8.4) and (8.5) are now stated.

$$\int \frac{1}{x}\,\mathrm{d}x = \ln |x|. \tag{8.8}$$

$$\int \mathrm{e}^x\,\mathrm{d}x = \mathrm{e}^x. \tag{8.9}$$

The *modulus function* $|x|$, defined in problem 6 of Chapter 0, is used in equation (8.8) to cater for situations when x is *negative*. It is not incorrect to omit the modulus notation, but its inclusion is useful when calculating certain definite integrals.

For $\int (1/x)\,\mathrm{d}x$, $x < 0$, let $y = -x$, $\mathrm{d}y = -\mathrm{d}x$. Therefore,

$$\begin{aligned}\int \frac{1}{x}\,\mathrm{d}x &= \int \frac{1}{-y}(-\mathrm{d}y)\\ &= \int \frac{1}{y}\,\mathrm{d}y\\ &= \ln y + C\end{aligned}$$

since $y > 0$.

$$\begin{aligned}\ln y + C &= \ln(-x) + C\\ &= \ln |x| + C.\end{aligned}$$

It should be noted that equation (8.8) completes the rule for integrating x^n, equation (4.3), which did not apply for the case $n = -1$.

Example 8.1

Differentiate the following functions with respect to x.

(i) e^{2x}.
(ii) e^{-x^2}.
(iii) $\ln(\cos x)$.
(iv) $\ln(x^2 \cos x)$.

(i) From equation (8.7b),

$$\begin{aligned}\frac{\mathrm{d}}{\mathrm{d}x}(\mathrm{e}^{2x}) &= \mathrm{e}^{2x}\frac{\mathrm{d}}{\mathrm{d}x}(2x)\\ &= 2\mathrm{e}^{2x}.\end{aligned}$$

(ii) From equation (8.7b),

$$\frac{d}{dx}(e^{-x^2}) = e^{-x^2}\frac{d}{dx}(-x^2)$$
$$= -2x\, e^{-x^2}.$$

(iii) From equation (8.6b),

$$\frac{d}{dx}\{\ln(\cos x)\} = \frac{1}{\cos x}\frac{d}{dx}(\cos x)$$
$$= -\frac{\sin x}{\cos x}$$
$$= -\tan x.$$

(iv) The function $\ln(x^2 \cos x)$ may be simplified using the laws of logarithms.

$$\ln(x^2 \cos x) = \ln(x^2) + \ln(\cos x)$$
$$= 2\ln x + \ln(\cos x).$$

Therefore,

$$\frac{d}{dx}\{\ln(x^2 \cos x)\} = \frac{2}{x} - \frac{\sin x}{\cos x}$$
$$= \frac{2}{x} - \tan x.$$

Example 8.2

Carry out the following integrations.

(i) $\int e^{3x}\, dx.$

(ii) $\int x\, e^{3x^2}\, dx.$

(iii) $\int \frac{dx}{3x-1}.$

(iv) $\int \frac{x-3}{x+3}\, dx.$

(v) $\int_{\pi/6}^{\pi/2} \cot\theta\, d\theta.$

(vi) $\int_0^1 \frac{t^2}{3+t^3}\, dt.$

(i) $$\int e^{3x}\,dx = \tfrac{1}{3}e^{3x} + C,$$

since

$$\frac{d}{dx}(e^{3x}) = 3e^{3x}.$$

(ii) Substitute $u = 3x^2$, $du = 6x\,dx$, i.e. $x\,dx = \tfrac{1}{6}\,du$.

$$\int x\,e^{3x^2}\,dx = \int e^u\,\tfrac{1}{6}\,du$$

$$= \tfrac{1}{6}\int e^u\,du$$

From equation (8.9),

$$\tfrac{1}{6}\int e^u\,du = \tfrac{1}{6}e^u + C$$

$$= \tfrac{1}{6}e^{3x^2} + C.$$

(iii) $$\int \frac{dx}{3x-1}$$

is just another way of writing

$$\int \frac{1}{3x-1}\,dx.$$

Substitute $u = 3x - 1$, $du = 3\,dx$, i.e. $dx = \tfrac{1}{3}\,du$.

$$\int \frac{dx}{3x-1} = \int \frac{\tfrac{1}{3}du}{u}$$

$$= \tfrac{1}{3}\int \frac{du}{u}.$$

From equation (8.8),

$$\tfrac{1}{3}\int \frac{du}{u} = \tfrac{1}{3}\ln|u| + C$$

$$= \tfrac{1}{3}\ln|3x-1| + C.$$

(iv) Substitute $u = x + 3$, $du = dx$, $x - 3 = u - 6$. This substitution allows the rational function

$$\frac{x-3}{x+3}$$

to be expressed in a form with a *single-term* denominator; simple division can then be carried out.

$$\int \frac{x-3}{x+3}\,dx = \int \frac{u-6}{u}\,du$$
$$= \int \left(1 - \frac{6}{u}\right) du$$
$$= u - 6 \ln |u| + C$$
$$= x + 3 - 6 \ln |x+3| + C,$$

which can be written as $x - 6 \ln |x + 3| + C$ since C is arbitrary.

(v) $$\cot\theta = \frac{\cos\theta}{\sin\theta}.$$

Substitute $u = \sin\theta$, $du = \cos\theta\,d\theta$. When $\theta = \pi/6$, $u = \frac{1}{2}$; when $\theta = \pi/2$, $u = 1$.

$$\int_{\pi/6}^{\pi/2} \cot\theta\,d\theta = \int_{1/2}^{1} \frac{du}{u}$$
$$= [\ln |u|]_{1/2}^{1}$$
$$= \ln 1 - \ln \tfrac{1}{2}$$
$$= \ln 2$$

(since $\ln(2^{-1}) = -\ln 2$).

(vi) Substitute $u = 3 + t^3$, $du = 3t^2\,dt$, i.e. $t^2\,dt = \frac{1}{3}\,du$. When $t = 0$, $u = 3$; when $t = 1$, $u = 4$.

$$\int_0^1 \frac{t^2}{3+t^3}\,dt = \int_3^4 \frac{\frac{1}{3}du}{u}$$
$$= \tfrac{1}{3} \int_3^4 \frac{du}{u}$$
$$= \tfrac{1}{3}\,[\ln |u|]_3^4$$
$$= \tfrac{1}{3}(\ln 4 - \ln 3)$$
$$= \tfrac{1}{3} \ln\left(\tfrac{4}{3}\right).$$

Example 8.3

Prove that $\int \sec x\,dx = \ln(\sec x + \tan x) + C$.

We simply differentiate the right-hand side to obtain $\sec x$.

$$\frac{\mathrm{d}}{\mathrm{d}x}\{\ln(\sec x + \tan x)\} = \frac{\sec x \tan x + \sec^2 x}{\sec x + \tan x}$$

$$= \frac{\sec x\,(\sec x + \tan x)}{\sec x + \tan x}$$

$$= \sec x.$$

Therefore,

$$\int \sec x\,\mathrm{d}x = \ln(\sec x + \tan x) + C.$$

Example 8.4

Calculate the value of $\mathrm{d}y/\mathrm{d}x$ when $x = 0$ for the function

$$y = \frac{\mathrm{e}^{4x}\cos^2 x}{(2+x)^4}.$$

We apply the technique of logarithmic differentiation to simplify the given function and so to avoid use of the product and quotient rules.

$$\ln y = \ln(\mathrm{e}^{4x}\cos^2 x) - \ln(2+x)^4$$

$$= \ln(\mathrm{e}^{4x}) + \ln(\cos^2 x) - 4\ln(2+x),$$

i.e.

$$\ln y = 4x + 2\ln(\cos x) - 4\ln(2+x).$$

Differentiating the equation with respect to x,

$$\frac{1}{y}\frac{\mathrm{d}y}{\mathrm{d}x} = 4 + 2\frac{(-\sin x)}{\cos x} - 4\frac{1}{2+x}$$

and

$$\frac{\mathrm{d}y}{\mathrm{d}x} = y\left(4 - 2\tan x - \frac{4}{2+x}\right).$$

When $x = 0, y = \frac{1}{16}$ and $\mathrm{d}y/\mathrm{d}x = \frac{1}{16}(4 - 0 - 2) = \frac{1}{8}$.

Example 8.5

Use logarithmic differentiation to obtain $\mathrm{d}y/\mathrm{d}x$ when $y = a^x$, where a is a positive constant.

$$y = a^x.$$

From equation (8.3),

$$\ln y = x\ln a.$$

$$\frac{1}{y}\frac{dy}{dx} = \ln a.$$

Therefore,

$$\frac{dy}{dx} = y \ln a$$

$$= a^x \ln a.$$

Problems

1. Differentiate the following functions with respect to x.
   (i) $e^{\sqrt{x}}$.
   (ii) $e^{\sin^2 x}$.
   (iii) $e^{2x} \ln x$.
   (iv) $\ln 2x + 2 \ln x$.
   (v) $\ln(\sqrt{x^2 + 3x + 1})$.
   (vi) $\ln\left(\frac{1-x}{1+x}\right)$.
   (vii) $\frac{e^{2x}}{1+x^2}$.

2. Obtain the following derivatives.
   (i) $\frac{d}{dt}\{\ln(e^t\, t^2 \sqrt{1+t^2})\}$.
   (ii) $\frac{d}{dy}\{\ln(1 + e^{2y})\}$.
   (iii) $\frac{d}{du}\left\{\ln\left(\frac{\sec u}{u}\right)\right\}$.

3. (i) Show that $y = (3 + 2t)\, e^{-t}$ is a solution of the equation

$$\frac{d^2y}{dt^2} + 2\frac{dy}{dt} + y = 0.$$

   (ii) Show that $V = A\, e^{-x} + B\, e^{-2x} + 2x - 3$ satisfies the equation

$$\frac{d^2 V}{dt^2} + 3\frac{dV}{dt} + 2V = 4x$$

   for all values of the constants A and B.

4. Carry out the following integrations.

(i) $\int e^{2x-3}\, dx$.

(ii) $\int \cos x\, e^{2\sin x}\, dx$.

(iii) $\int (e^{x/3} - e^{-x/3})^2\, dx$.

(iv) $\int \frac{s+2}{s^2+4s+5}\, ds$.

(v) $\int \frac{e^x}{1+2e^x}\, dx$.

(vi) $\int_{-2}^{-1/2} \frac{dt}{1-4t}$.

(vii) $\int_0^1 \frac{y}{y+1}\, dy$.

(viii) $\int \frac{dx}{(2x+3)^2}$.

5. $A(t)$ is the area under the curve

$$y = \frac{1}{x}$$

between $x = 1$ and $x = t$, $t > 1$. $V(t)$ is the volume generated when $A(t)$ is rotated about the x axis. Determine the functions $A(t)$ and $V(t)$. Show that $\lim_{t\to\infty} \{A(t)\}$ does not exist, but that $\lim_{t\to\infty} \{V(t)\} = \pi$.

6. Show that

$$\int \frac{dx}{\sqrt{4x^2 - 4x + 10}} = \tfrac{1}{2} \ln \{(2x-1) + \sqrt{4x^2 - 4x + 10}\} + C.$$

7. Use logarithmic differentiation to obtain the first derivative with respect to x of the following functions.

(i) $y = (1-2x)^3 (3+2x)^4$.

(ii) $y = \sqrt{x(x+1)}$.

(iii) $y = x^3 e^{2x} \sin^2 5x$.

(iv) $y = x^x$.

8. Evaluate the following, correct to two decimal places.
   (i) The gradient of the curve $y = 10^x$ at the point (0, 1).
   (ii) $\int_0^2 10^x \, dx$.

9. Use logarithmic differentiation to derive the product rule and the quotient rule.

10. Find the equation of the tangent to the curve

$$y = \frac{2\sqrt{3 + x^2}}{(1 + x^2)x^x}$$

at the point (1, 2).

11. The concentration of a drug in the bloodstream t h after injection is given by the formula

$$C = A\,e^{-at} + B\,e^{-bt},$$

where A, B, a and b are constants. Write down an expression for the rate of change of the concentration when $t = 10$.

12. The rate of growth of the number of cells in a culture is given by dx/dt where

$$x = \frac{x_M}{1 + a\,e^{-kt}},$$

and x_M, a and k are constants. Show that the rate of growth is maximum when $t = (1/k)\ln a$.

Answers

1. (i) $\dfrac{e^{\sqrt{x}}}{2\sqrt{x}}$.

   (ii) $2 \sin x \cos x \, e^{\sin^2 x}$.

   (iii) $e^{2x}\left(2 \ln x + \dfrac{1}{x}\right)$.

   (iv) $\dfrac{3}{x}$.

   (v) $\dfrac{2x + 3}{2(x^2 + 3x + 1)}$.

   (vi) $-\dfrac{2}{1 - x^2}$.

(vii) $\dfrac{2\,e^{2x}(x^2 - x + 1)}{(1 + x^2)^2}$.

2. (i) $1 + \dfrac{2}{t} + \dfrac{t}{1 + t^2}$.

(ii) $\dfrac{2\,e^{2y}}{1 + e^{2y}}$.

(iii) $\tan u - \dfrac{1}{u}$.

4. (i) $\frac{1}{2}\,e^{2x-3} + C$.
(ii) $\frac{1}{2}\,e^{2\sin x} + C$.
(iii) $\frac{3}{2}\,e^{2x/3} - 2x - \frac{3}{2}\,e^{-2x/3} + C$.
(iv) $\frac{1}{2}\ln(s^2 + 4s + 5) + C$.
(v) $\frac{1}{2}\ln(1 + 2e^x) + C$.
(vi) $\frac{1}{4}\ln 3$.
(vii) $1 - \ln 2$.

(viii) $-\dfrac{1}{2(2x + 3)} + C$.

5. $A(t) = \ln t$, $V(t) = \pi(1 - t^{-1})$.

7. (i) $2y\left(\dfrac{4}{3 + 2x} - \dfrac{3}{1 - 2x}\right)$.

(ii) $\frac{1}{2}y\left(\dfrac{1}{x} + \dfrac{1}{x + 1}\right)$.

(iii) $y\left(\dfrac{3}{x} + 2 + 10\cot 5x\right)$.

(iv) $x^x(1 + \ln x)$.

8. (i) 2.30.
(ii) 42.99.

10. $7x + 2y - 11 = 0$.

11. $-(aA\,e^{-10a} + bB\,e^{-10b})$.

8.3 GROWTH AND DECAY PROBLEMS

Many practical situations are modelled approximately by an equation in which the rate of change in a certain variable is directly proportional to the variable itself. Denoting the variable by y, the equation of proportionality can be written

$$\frac{dy}{dt} = ky \tag{8.10}$$

where k is a constant.

If we are given additionally that y takes the value y_0 when $t = 0$, then it is easy to verify that the solution of equation (8.10) is

$$y = y_0\, e^{kt}. \tag{8.11}$$

(Simply differentiate y with respect to t in equation (8.11) and show that equation (8.10) is satisfied. The condition that $y = y_0$ when $t = 0$ is confirmed by setting t to the value zero in equation (8.11).) If $k > 0$, then $y \to \infty$ as $t \to \infty$, i.e. y *grows* (e.g. increasing population). If $k < 0$, then $y \to 0$ as $t \to \infty$, i.e. y *decays* (e.g. radioactive decay).

Fig. 8.2 shows typical growth and decay graphs.

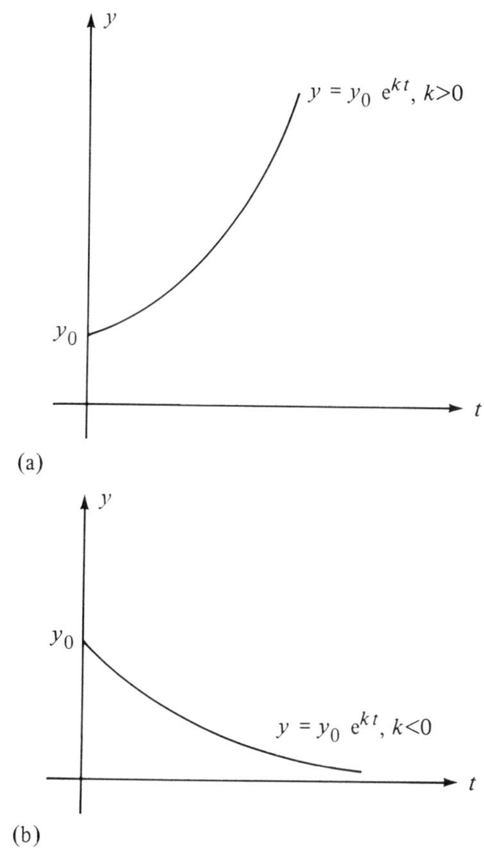

Fig. 8.2 (a) A growth curve; (b) a decay curve.

Example 8.6

Some material heated to 100°C is placed in a room in which the temperature, assumed constant, is 20°C. The rate of change in temperature difference is assumed to be proportional to the temperature difference. (This is known as Newton's law of cooling.)

(i) If the temperature of the material is 80°C after 4 min, find the constant of proportionality.
(ii) What will be the temperature of the material 20 min after cooling has begun?.
(iii) How long will the material take to cool to 30°C?

Let the temperature difference be y°C. Therefore, $\mathrm{d}y/\mathrm{d}t = ky$, where k min^{-1} is the constant of proportionality.

We are given that $y = 80$°C when $t = 0$, the time when the material is introduced into the constant-temperature room. Applying equation (8.11), we have $y = 80\,\mathrm{e}^{kt}$.

(i) When $t = 4$ min, $y = 80 - 20°\mathrm{C} = 60°\mathrm{C}$. Therefore,

$$60 = 80\,\mathrm{e}^{4k}$$

i.e.

$$\mathrm{e}^{4k} = 0.75$$

i.e.

$$k = \tfrac{1}{4}\ln 0.75$$
$$\approx -0.0719\ \mathrm{min}^{-1}.$$

(ii) $y = 80\,\mathrm{e}^{-0.0719t}$.
When $t = 20$, $y = 80\,\mathrm{e}^{-1.438} = 19°\mathrm{C}$ approximately. Therefore the temperature of the material is about 39°C.

(iii) When the temperature of the material is 30 °C, $y = 30 - 20\,°\mathrm{C} = 10\,°\mathrm{C}$ and $10 = 80\,\mathrm{e}^{-0.0719t}$ at the required time, i.e.

$$0.0719t = \ln 8$$

and

$$t = \frac{2.0794}{0.0719} \approx 28.9\ \mathrm{min}.$$

Example 8.7

The rate of increase in the number of bacteria in a culture is proportional to the number present. If there were initially 200 bacteria and this number increased to 320 after 10 min, find the number of bacteria present after 20 min.

Let N be the number of bacteria. Therefore, $\mathrm{d}N/\mathrm{d}t = kN$ bacteria per minute, and $N = 200$ when $t = 0$, i.e. $N = 200\,\mathrm{e}^{kt}$.

Also $N = 320$ when $t = 10$, i.e. $320 = 200\,\mathrm{e}^{10k}$, i.e. $\mathrm{e}^{10k} = 1.6$ and

$$k = \frac{\ln(1.6)}{10} \approx 0.047 \text{ min}^{-1}.$$

Therefore,

$$N = 200 \exp\left\{\frac{\ln(1.6)}{10}\, t\right\}.$$

When $t = 20$,

$$\begin{aligned} N &= 200\, e^{2 \ln(1.6)} \\ &= 200\, e^{\ln\{(1.6)^2\}} \\ &= 200 \times (1.6)^2 \\ &= 512, \end{aligned}$$

i.e. there are 512 bacteria present after 20 min.

(This result can be obtained by noting that the number of bacteria increased by a factor of $320/200 = 1.6$ in 10 min. Since the growth is *exponential*, the number will increase by this factor in the *next* 10 min. Therefore, after 20 min, the number of bacteria is $200 \times (1.6)^2 = 512$.)

Problems

1. The rate of increase in a certain population is proportional to the size of population. If the population doubles in 20 years, how long does it take to increase by a factor of 5?

2. The rate at which certain machinery depreciates is proportional to the value of the machinery. If a machine cost £30 000 when new and is only worth £15 000 after 5 years, find its value after the following periods of time.
   (i) 10 years.
   (ii) 6 years.
   (iii) 18 years.

3. The rate of decay of a radioactive element is proportional to the mass of the element. When the mass is 0.8×10^{-2} g, the rate of decay is 0.1×10^{-5} g/s. Find how long it takes the mass to be reduced from 0.8×10^{-2} g to 0.32×10^{-2} g.

4. The half-life of a radioactive element is the time taken for its mass to be halved. If the half-life of radium is 1690 years, find the time it would take for 16 g of radium to reduce to 2 g.

5. The electricity required by a large company increases exponentially, doubling every 10 years. Express the electricity requirement in 6 years' time as a factor of the present requirement.

6. In a certain electrical circuit the current i satisfies the equation

$$L\frac{\mathrm{d}i}{\mathrm{d}t} + Ri = V, \qquad i(0) = 0.$$

Making the substitution

$$i = I + \frac{V}{R},$$

show that

$$i = \frac{V}{R}(1 - \mathrm{e}^{-Rt/L}).$$

7. The number of organisms of a certain type in a population grows exponentially at a rate $\mathrm{d}N/\mathrm{d}t = (\ln 2^{100})\,\mathrm{e}^{t \ln 2}$ members per day. The number in the population after T days is given by

$$\lim_{s\to\infty} \int_{-s}^{T} \frac{\mathrm{d}N}{\mathrm{d}t}\,\mathrm{d}t.$$

Calculate the number of organisms present in the population
(i) when $T = 0$.
(ii) after 5 days.

Answers

1. 45 years approximately.
2. (i) £7500.
   (ii) £13 062.
   (iii) £2476.
3. 122.2 min.
4. 5070 years.
5. The factor is 1.52 approximately.
7. (i) 100
   (ii) 3200.

8.4 HYPERBOLIC FUNCTIONS

Hyperbolic functions have applications in some branches of engineering and a brief description of their properties is given in this section.

We define the **hyperbolic sine** and **hyperbolic cosine** functions as follows:

$$\sinh x = \tfrac{1}{2}(\mathrm{e}^{x} - \mathrm{e}^{-x}), \tag{8.12}$$

$$\cosh x = \tfrac{1}{2}(\mathrm{e}^{x} + \mathrm{e}^{-x}). \tag{8.13}$$

Additionally we define four remaining hyperbolic functions:

$$\tanh x = \frac{\sinh x}{\cosh x} = \frac{e^{2x} - 1}{e^{2x} + 1}, \tag{8.14}$$

$$\operatorname{cosech} x = \frac{1}{\sinh x}, \tag{8.15}$$

$$\operatorname{sech} x = \frac{1}{\cosh x}, \tag{8.16}$$

$$\coth x = \frac{1}{\tanh x}. \tag{8.17}$$

The graphs of the functions $y = \sinh x$, $y = \cosh x$ and $y = \tanh x$ are shown in Fig. 8.3.
Note the *ranges* for these functions:

$$-\infty < \sinh x < \infty,$$

$$\cosh x \geqslant 1,$$

$$-1 < \tanh x < 1.$$

The hyperbolic functions have certain properties which resemble those of the six trigonometric functions.

For example, the identity $\sinh 2x = 2 \sinh x \cosh x$ may be proved directly from the definitions of the functions $\sinh x$ and $\cosh x$ and compares with the trigonometric identity $\sin 2x = 2 \sin x \cos x$.

Also the identity

$$\cosh^2 x - \sinh^2 x = 1$$

is similar to $\cos^2 x + \sin^2 x = 1$.

For every trigonometric identity given in section 0.7, there is a corresponding hyperbolic identity.

The following rule (known as Osborn's rule) shows how an identity for hyperbolic functions may be written down when the corresponding trigonometric identity is given.

Osborn's Rule

(i) Change the trigonometric functions into the corresponding hyperbolic functions.
(ii) Change the *sign* wherever there appears a product (or implied product) of two sinh functions.

For example,

$$\cos(x + y) = \cos x \cos y - \sin x \sin y$$

becomes

$$\cosh(x + y) = \cosh x \cosh y + \sinh x \sinh y.$$

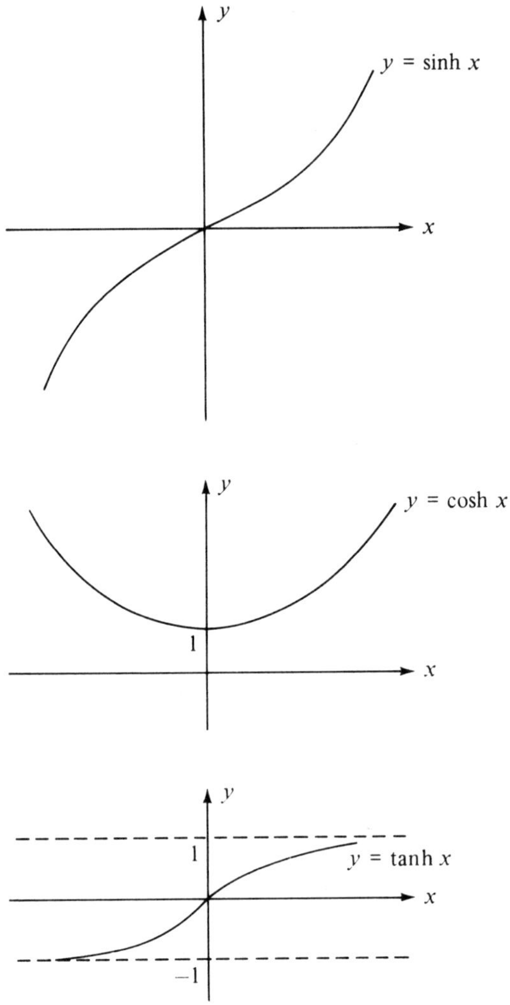

Fig. 8.3 Graphs of hyperbolic functions.

Also

$$1 + \tan^2 x = \sec^2 x$$

becomes

$$1 - \tanh^2 x = \sec^2 x.$$

$$\tanh^2 x = \frac{\sinh^2 x}{\cosh^2 x}$$

is an *implied* product of two sinh functions.

It should be understood that Osborn's rule does not constitute a *proof* for hyperbolic identities. All identities may be proved using the definitions of the six functions.

The derivatives of the six hyperbolic functions may be obtained from the definitions (8.12)–(8.17) and are listed below (note the similarity to the rules (2.3)–(2.8) for the trigonometric functions):

$$\frac{\mathrm{d}}{\mathrm{d}x}(\sinh x) = \cosh x, \tag{8.18}$$

$$\frac{\mathrm{d}}{\mathrm{d}x}(\cosh x) = \sinh x, \tag{8.19}$$

$$\frac{\mathrm{d}}{\mathrm{d}x}(\tanh x) = \mathrm{sech}^2\, x, \tag{8.20}$$

$$\frac{\mathrm{d}}{\mathrm{d}x}(\mathrm{cosech}\, x) = -\,\mathrm{cosech}\, x \coth x, \tag{8.21}$$

$$\frac{\mathrm{d}}{\mathrm{d}x}(\mathrm{sech}\, x) = -\,\mathrm{sech}\, x \tanh x, \tag{8.22}$$

$$\frac{\mathrm{d}}{\mathrm{d}x}(\coth x) = -\,\mathrm{cosech}^2\, x. \tag{8.23}$$

Six standard integrals correspond to the above rules. In particular, we have

$$\int \cosh x \, \mathrm{d}x = \sinh x, \tag{8.24}$$

$$\int \sinh x \, \mathrm{d}x = \cosh x. \tag{8.25}$$

Example 8.8

Use the definitions (8.12) and (8.13) to prove the following identities.

(i) $\sinh 2x = 2 \sinh x \cosh x$.

(ii) $\cosh^2 x - \sinh^2 x = 1$.

(i) $$2 \sinh x \cosh x = 2\left(\frac{\mathrm{e}^x - \mathrm{e}^{-x}}{2}\right)\left(\frac{\mathrm{e}^x + \mathrm{e}^{-x}}{2}\right) = \frac{\mathrm{e}^{2x} - \mathrm{e}^{-2x}}{2}.$$

From equation (8.12),

$$\frac{\mathrm{e}^{2x} - \mathrm{e}^{-2x}}{2} = \sinh 2x.$$

(ii) We note from the definitions of $\sinh x$ and $\cosh x$ that

$$\cosh x + \sinh x = e^x,$$

$$\cosh x - \sinh x = e^{-x}.$$

Therefore,

$$(\cosh x + \sinh x)(\cosh x - \sinh x) = e^x e^{-x},$$

i.e.

$$\cosh^2 x - \sinh^2 x = 1.$$

Example 8.9

Use Osborn's rule to obtain an identity expressing $\cosh 2x$ in terms of $\sinh x$. Hence solve the equation

$$\cosh 2x - 13 \sinh x + 19 = 0.$$

Equation (0.9) is

$$\sin^2 x = \tfrac{1}{2}(1 - \cos 2x).$$

Applying Osborn's rule,

$$-\sinh^2 x = \tfrac{1}{2}(1 - \cosh 2x),$$

i.e.

$$\cosh 2x = 1 + 2 \sinh^2 x.$$

The equation now becomes

$$1 + 2 \sinh^2 x - 13 \sinh x + 19 = 0,$$

i.e.

$$2 \sinh^2 x - 13 \sinh x + 20 = 0.$$

Factorizing gives

$$(2 \sinh x - 5)(\sinh x - 4) = 0$$

and

$$\sinh x = \tfrac{5}{2} \quad \text{or} \quad \sinh x = 4.$$

Most scientific calculators include facilities for the evaluation of inverse hyperbolic functions. Alternatively, we may proceed by applying definition (8.12).

$$e^x - e^{-x} = 5 \quad \text{or} \quad e^x - e^{-x} = 8,$$

i.e.

$$(e^x)^2 - 5(e^x) - 1 = 0 \quad \text{or} \quad (e^x)^2 - 8(e^x) - 1 = 0$$

and

$$e^x = \frac{5 \pm \sqrt{29}}{2} \quad \text{or} \quad e^x = \frac{8 \pm \sqrt{68}}{2}.$$

Rejecting the negative signs since $e^x > 0$ always,

$$x = \ln\left(\frac{5 + \sqrt{29}}{2}\right) \quad \text{or} \quad x = \ln(4 + \sqrt{17}).$$

Therefore the solutions are $x \approx 1.647$ and $x \approx 2.095$.

Example 8.10

Differentiate the following functions with respect to x.

(i) $\cosh 5x$.

(ii) $\sinh^2 3x$.

(iii) $\ln\left\{\tanh\left(\frac{x}{2}\right)\right\}$.

(i) Let $y = \cosh 5x = \cosh u$, where $u = 5x$.

$$\begin{aligned}\frac{dy}{dx} &= \frac{dy}{du}\frac{du}{dx} \\ &= (\sinh u)(5) \\ &= 5 \sinh 5x.\end{aligned}$$

(ii) Let $y = \sinh^2 3x = u^2$, where $u = \sinh 3x$.

$$\begin{aligned}\frac{dy}{dx} &= \frac{dy}{du}\frac{du}{dx} \\ &= (2u)(3 \cosh 3x) \\ &= 6 \sinh 3x \cosh 3x.\end{aligned}$$

(iii)
$$\begin{aligned}\frac{d}{dx}\left[\ln\left\{\tanh\left(\frac{x}{2}\right)\right\}\right] &= \frac{1}{\tanh(x/2)}\frac{d}{dx}\left\{\tanh\left(\frac{x}{2}\right)\right\} \\ &= \frac{1}{\tanh(x/2)}\,\tfrac{1}{2}\operatorname{sech}^2\left(\frac{x}{2}\right) \\ &= \frac{1}{2\sinh(x/2)\cosh(x/2)} \\ &= \frac{1}{\sinh x} \\ &= \operatorname{cosech} x.\end{aligned}$$

Example 8.11
Evaluate the following integrals.

(i) $\displaystyle\int_0^{0.2} \sinh 3x \, dx.$

(i) $\displaystyle\int_0^{1} x \cosh(1 + x^2) \, dx.$

(iii) $\displaystyle\int_0^{\ln 2} \tanh t \, dt.$

(i) Let $u = 3x$, $du = 3\, dx$ and $dx = \frac{1}{3}\, du$. When $x = 0$, $u = 0$; when $x = 0.2$, $u = 0.6$.

$$\begin{aligned}\int_0^{0.2} \sinh 3x \, dx &= \tfrac{1}{3} \int_0^{0.6} \sinh u \, du \\ &= \tfrac{1}{3} [\cosh u]_0^{0.6} \\ &= \tfrac{1}{3} \{\cosh (0.6) - \cosh 0\} \\ &\approx \tfrac{1}{3} (1.1855 - 1) \\ &\approx 0.0618.\end{aligned}$$

(ii) Let $u = 1 + x^2$, $du = 2x\, dx$ and $x\, dx = \frac{1}{2}\, du$. When $x = 0$, $u = 1$; when $x = 1$, $u = 2$.

$$\begin{aligned}\int_0^{1} x \cosh(1 + x^2) \, dx &= \tfrac{1}{2} \int_1^{2} \cosh u \, du \\ &= \tfrac{1}{2} [\sinh u]_1^2 \\ &= \tfrac{1}{2} (\sinh 2 - \sinh 1) \\ &\approx \tfrac{1}{2} (3.6269 - 1.1752) \\ &\approx 1.226.\end{aligned}$$

(iii) $\displaystyle\int_0^{\ln 2} \tanh t \, dt = \int_0^{\ln 2} \frac{\sinh t}{\cosh t} \, dt.$

Let $u = \cosh t$, $du = \sinh t\, dt$. When $t = 0$, $u = 1$. When $t = \ln 2$,

$$u = \cosh(\ln 2) = \frac{2 + 2^{-1}}{2} = \tfrac{5}{4}.$$

$$\begin{aligned}\int_0^{\ln 2} \tanh t \, dt &= \int_1^{5/4} \frac{du}{u} \\ &= [\ln u]_1^{5/4} \\ &= \ln(1.25) - \ln 1 \\ &\approx 0.2231.\end{aligned}$$

Example 8.12

Find the function $\int \cosh x \cosh 2x \, dx$ by the following means.

(i) Using an identity.
(ii) Using the definition of $\cosh x$.

(i) $$\int \cosh x \cosh 2x \, dx = \tfrac{1}{2}\int (\cosh 3x + \cosh x) \, dx$$
$$= \tfrac{1}{2}\left(\tfrac{1}{3}\sinh 3x + \sinh x\right) + C$$
$$= \tfrac{1}{6}\sinh 3x + \tfrac{1}{2}\sinh x + C.$$

(ii) $$\int \cosh x \cosh 2x \, dx = \int \tfrac{1}{2}(e^x + e^{-x})\,\tfrac{1}{2}(e^{2x} + e^{-2x}) \, dx$$
$$= \tfrac{1}{4}\int (e^{3x} + e^{-3x} + e^x + e^{-x}) \, dx$$
$$= \tfrac{1}{4}\left(\tfrac{1}{3}e^{3x} - \tfrac{1}{3}e^{-3x} + e^x - e^{-x}\right) + C$$
$$= \tfrac{1}{6}\left\{\tfrac{1}{2}(e^{3x} - e^{-3x})\right\} + \tfrac{1}{2}\left\{\tfrac{1}{2}(e^x - e^{-x})\right\} + C$$
$$= \tfrac{1}{6}\sinh 3x + \tfrac{1}{2}\sinh x + C.$$

Problems

1. Use Osborn's rule to establish the identity $\cosh 2x = \cosh^2 x + \sinh^2 x$ and hence express $\cosh 2x$ in terms of $\cosh x$. Solve the equation

$$3\cosh^2 x - \cosh 2x - 5 = 0.$$

2. Use the definitions of $\sinh x$ and $\cosh x$
   (i) to prove the identity given in problem 1,
   (ii) to solve the equation $\sinh x - 2\cosh x + 2 = 0$.

3. Differentiate the following functions with respect to x.
   (i) $\cosh 3x \sinh 3x$.
   (ii) $\dfrac{\sinh 3x}{1 + \cosh 3x}$.
   (iii) $\mathrm{sech}^2\, 4x$.
   (iv) $\ln(x^2 \cosh 6x)$.
   (v) $2^{\sinh x}$.

4. Evaluate the following definite integrals.
   (i) $\displaystyle\int_3^6 \cosh\left(\frac{x}{12}\right) dx$.

(ii) $\displaystyle\int_0^{0.5} \frac{\sinh 2y}{\cosh^2 2y}\,\mathrm{d}y.$

(iii) $\displaystyle\int_0^{1} \cosh^2 x\,\mathrm{d}x.$

(iv) $\displaystyle\int_0^{0.1} \mathrm{e}^u \sinh 2u\,\mathrm{d}u.$

5. Show that

$$y = A\cosh nx + B\sinh nx$$

is a solution of the equation

$$\frac{\mathrm{d}^2y}{\mathrm{d}x^2} - n^2y = 0$$

for any values of the constants A, B and n.

6. A cable is suspended from two points in a horizontal line, of distance $2a$ apart. The cable takes the shape of the curve

$$y = c\cosh\left(\frac{x}{c}\right)$$

where c is a positive constant (Fig. 8.4). (This curve is called a **catenary**.) It is given that the total length of the cable is $2l$ where

$$l = c\sinh\left(\frac{a}{c}\right).$$

Show that the sag in the cable is $\sqrt{c^2 + l^2} - c$.

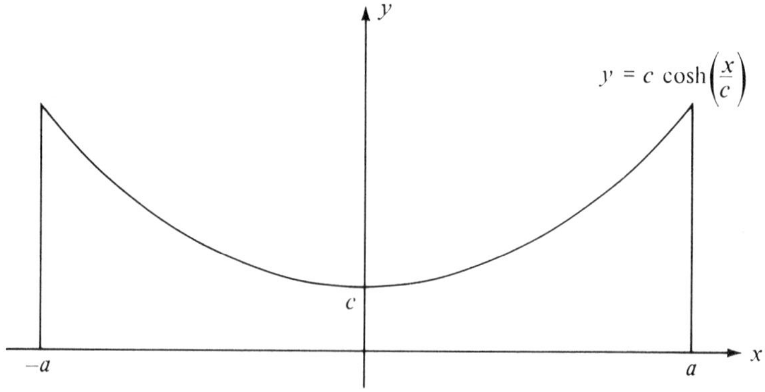

Fig. 8.4 The catenary.

7. Show that the radius of curvature of the curve

$$y = c \cosh\left(\frac{x}{c}\right)$$

at the point where $x = t$ is given by

$$\rho = c \cosh^2\left(\frac{t}{c}\right),$$

where c is a positive constant.

8. A body of mass m falls through the air from some great height. Assuming that the only resistance to motion is proportional to the square of its velocity the equation of motion is

$$m\frac{dv}{dt} = mg - kv^2,$$

where v is velocity, t is time, and g and k are constants. Show that the equation is satisfied by

$$v = \sqrt{\frac{mg}{k}} \tanh\left(\sqrt{\frac{kg}{m}}\, t\right).$$

Write down the terminal velocity of the body.

Answers

1. ± 1.317.
2. (ii) 1.0986, 0.
3. (i) $5 \cosh 3x \cosh 5x + 3 \sinh 3x \sinh 5x$.

   (ii) $\dfrac{3}{1 + \cosh 3x}$.

   (iii) $-8 \operatorname{sech}^2 4x \tanh 4x$.

   (iv) $\dfrac{2}{x} + 6 \tanh 6x$.

   (v) $\ln 2 \cosh x \, 2^{\sinh x}$.
4. (i) 3.222.
   (ii) 0.176.
   (iii) 1.067.
   (iv) 0.0107.

8. $\sqrt{\dfrac{mg}{k}}$.

Further Problems for Chapter 8

1. Given that $\ln(1 - T^2) = y + 2$, express t in terms of y.

2. Given that $\ln y = 3 \ln x + \ln(3 - x) - \ln 3$ express y as a polynomial in x.

3. Evaluate $f'(1)$ when $f(x) = \ln(1 - 3x)$.

4. Determine the gradient of the curve $y = \mathrm{e}^{4-x^2}$ at the point (2, 1).

5. Use logarithmic differentiation to obtain the value of $\mathrm{d}y/\mathrm{d}x$ at the point (1, 2) on the curve

$$y = \sqrt{\frac{3+x}{2-x}}.$$

6. Evaluate

$$\int_{-2}^{-1} \frac{\mathrm{d}x}{x}.$$

7. Evaluate

$$\int_{0}^{2} \frac{\mathrm{d}y}{1+4y}$$

correct to three decimal places.

8. Given that $f'(t) = 4\mathrm{e}^{-2t}$ and $f(0) = 2$ determine the function $f(t)$.

9. Show that the curve $y = 10\mathrm{e}^{-x^2}$ has a maximum stationary point when $x = 0$.

10. Determine

$$\frac{\mathrm{d}}{\mathrm{d}\theta}(3^{1-\theta}).$$

11. Given that

$$\frac{\mathrm{d}x}{\mathrm{d}t} = -2x$$

and that $x = 6$ when $t = 0$, write down the expression for x in terms of t.

12. Use the definitions of the functions $\cosh x$ and $\sinh x$ to prove the identity

$$\cosh(x + y) = \cosh x \cosh y + \sinh x \sinh y.$$

13. Solve for x the equation $\sinh x = 0.75$, giving your answer correct to four decimal places.

14. Determine

$$\frac{\mathrm{d}}{\mathrm{d}x}\{\cosh(1-2x)\}.$$

15. Evaluate

$$\int_0^{0.25} \sinh 4x \, \mathrm{d}x,$$

correct to four decimal places.

Answers

1. $T = \sqrt[3]{1 - \mathrm{e}^{y+2}}$.
2. $y = \frac{1}{3}(3x^3 - x^4)$.
3. 1.5.
4. -4.
5. 1.25.
6. $-\ln 2$.
7. 0.549.
8. $4 - 2\mathrm{e}^{-2t}$.
10. $-3^{1-\theta} \ln 3$.
11. $x = 6\,\mathrm{e}^{-2t}$.
13. 0.6931.
14. $-2 \sinh(1 - 2x)$.
15. 0.1358.

9

Inverse Trigonometric and Hyperbolic Functions

9.1 INTRODUCTION

Given a function $y = f(x)$, suppose that we are able to rearrange the variables so that x is expressed in terms of y, i.e. $x = g(y)$. The notation $f^{-1}(y)$ is often used instead of $g(y)$. $f^{-1}(y)$ is called the **inverse function** corresponding to the function $f(x)$. (Note that, subject to possible restrictions on the range of the inverse function, $f^{-1}\{f(x)\} = x$ and $f\{f^{-1}(y)\} = y$ but that f^{-1} *does not mean* $1/f$).

When describing an inverse function the choice of letter used for the independent variable is unimportant. It is common practice to refer to $f^{-1}(x)$ as the inverse function corresponding to $f(x)$. For example, in Chapter 8 we studied the inverse logarithmic function e^x, i.e. $f^{-1}(x) = e^x$ is the inverse function corresponding to $f(x) = \ln x$. In other words, we can write

$$e^{\ln x} = \ln e^x = x.$$

Similarly $f^{-1}(x) = \sqrt[3]{x}$ is the inverse function corresponding to $f(x) = x^3$ since

$$\sqrt[3]{x^3} = (\sqrt[3]{x})^3 = x.$$

In this chapter the inverse functions corresponding to some of the trigonometric and hyperbolic functions are considered, together with their derivatives and the important corresponding integrals. Only the inverse sine function $\sin^{-1} x$ is studied in detail; the properties of the other functions are listed and proofs are left as exercises for the reader.

9.2 DEFINITIONS OF THE INVERSE FUNCTIONS

We define the **inverse sine function,** $y = \sin^{-1} x$, as follows: $y = \sin^{-1} x$ is equivalent to $x = \sin y$, $-\pi/2 \leqslant y \leqslant \pi/2$. In words, y is *the number whose sine is x*, where 'number' here refers to the non-dimensional angle (i.e. *radian* measure) in the closed interval between $-\pi/2$ and $\pi/2$.

The reason for the restriction on the range of the function $y = \sin^{-1} x$ is clear when we consider its graph. To draw the graph of $y = \sin^{-1} x$, we first draw $x = \sin y$, (Fig. 9.1). The range is restricted to $-\pi/2 \leqslant y \leqslant \pi/2$ to ensure that the function $y = \sin^{-1} x$ is a *single-valued* function of x. Note that the *domain* of the function $y = \sin^{-1} x$ is $-1 \leqslant x \leqslant 1$.

(An alternative notation for $\sin^{-1} x$ is $\arcsin x$.)

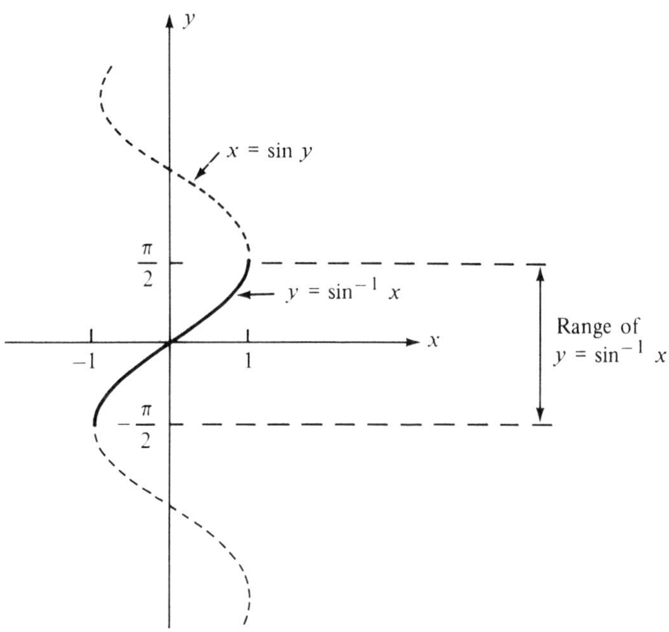

Fig. 9.1

Other useful inverse functions are defined below. In each case the domain is given and, where appropriate, range restrictions are stated. These restrictions are chosen to ensure that the functions are single valued. The reader should sketch the graph of each inverse function using the technique described for the function $y = \sin^{-1} x$. The necessity for the range restrictions may then be confirmed.

(i) $y = \cos^{-1} x$, $-1 \leqslant x \leqslant 1$, is such that $x = \cos y$, $0 \leqslant y \leqslant \pi$.
(ii) $y = \tan^{-1} x$, $-\infty < x < \infty$, is such that $x = \tan y$, $-\pi/2 < y < \pi/2$.
(iii) $y = \sinh^{-1} x$, $-\infty < x < \infty$, is such that $x = \sinh y$.

(iv) $y = \cosh^{-1} x$, $x \geqslant 1$, is such that $x = \cosh y$, $y \geqslant 0$.
(v) $y = \tanh^{-1} x$, $-1 < x < 1$, is such that $x = \tanh y$.

Most scientific calculators are able to evaluate the six inverse functions defined in this section. However, some calculators do not possess a hyperbolic function facility and the following logarithmic forms for inverse hyperbolic functions may be required:

$$\sinh^{-1} x = \ln(x + \sqrt{x^2 + 1}), \tag{9.1}$$

$$\cosh^{-1} x = \ln(x + \sqrt{x^2 - 1}), \qquad x \geqslant 1, \tag{9.2}$$

$$\tanh^{-1} x = \tfrac{1}{2} \ln\left(\frac{1+x}{1-x}\right), \qquad |x| < 1. \tag{9.3}$$

Equation (9.1) is derived in Example 9.2.

Example 9.1
Without using a calculator, write down the exact value of the following.
(i) $\cos^{-1}(-\tfrac{1}{2})$.
(ii) $\tan^{-1}(-\sqrt{3})$.
(iii) $\sinh^{-1}(\tfrac{3}{4})$.

(i) $y = \cos^{-1}(-\tfrac{1}{2})$ is the number such that $\cos y = -\tfrac{1}{2}$, $0 \leqslant y \leqslant \pi$, i.e. $y = 2\pi/3$.
(ii) $y = \tan^{-1}(-\sqrt{3})$ is the number such that $\tan y = -\sqrt{3}$, $-\pi/2 < y < \pi/2$, i.e. $y = -\pi/3$.
(iii) Using the logarithmic form, equation (9.1),

$$\sinh^{-1}(\tfrac{3}{4}) = \ln(\tfrac{3}{4} + \sqrt{\tfrac{9}{16} + 1})$$
$$= \ln 2.$$

Example 9.2
Derive the logarithmic form

$$\sinh^{-1} x = \ln(x + \sqrt{x^2 + 1}).$$

Let $y = \sinh^{-1} x$, i.e. $x = \sinh y$. Therefore,

$$x = \tfrac{1}{2}(e^y - e^{-y}),$$

i.e.

$$(e^y)^2 - 2x(e^y) - 1 = 0$$

and

$$e^y = \frac{2x \pm \sqrt{4x^2 + 4}}{2}$$
$$= x \pm \sqrt{x^2 + 1}.$$

We reject the negative sign since $e^y > 0$ for all y. Therefore,

$$e^y = x + \sqrt{x^2 + 1}$$

and

$$\begin{aligned} y &= \sinh^{-1} x \\ &= \ln(x + \sqrt{x^2 + 1}). \end{aligned}$$

Problems

1. Write down the exact value of
   (i) $\sin^{-1}(-1)$.
   (ii) $\cos^{-1}\left(\dfrac{\sqrt{3}}{2}\right)$.
   (iii) $\cos^{-1}\left(-\dfrac{\sqrt{3}}{2}\right)$.
   (iv) $\tan^{-1}(-1)$.
   (v) $\cosh^{-1}(\frac{5}{3})$.
   (iv) $\tanh^{-1}(0.5)$.

2. Without using a calculator, verify that

$$\sin^{-1}(\tfrac{3}{5}) + \sin^{-1}(\tfrac{4}{5}) = \frac{\pi}{2}.$$

3. Show that $\sin(2\cos^{-1} x) = 2x\sqrt{1-x^2}$.

4. Derive the following logarithmic forms.
   (i) $\cosh^{-1} x = \ln(x + \sqrt{x^2 - 1}), \quad x \geqslant 1,$
   (ii) $\tanh^{-1} x = \frac{1}{2} \ln\left(\dfrac{1+x}{1-x}\right), \quad |x| < 1.$

5. Show that, for $|x| < 1$ and $|y| < 1$,

$$\tan^{-1} x + \tan^{-1} y = \tan^{-1}\left(\frac{x+y}{1-xy}\right).$$

(Hint: let $A = \tan^{-1} x$, $B = \tan^{-1} y$ and use the formula

$$\tan(A+B) = \frac{\tan A + \tan B}{1 - \tan A \tan B}$$

to do this.)

Answers

1. (i) $-\dfrac{\pi}{2}$.

(ii) $\dfrac{\pi}{6}$.

(iii) $\dfrac{5\pi}{6}$.

(iv) $-\dfrac{\pi}{4}$.

(v) $\ln 3$.

(vi) $\frac{1}{2}\ln 3$.

9.3 DERIVATIVES AND INTEGRALS

The rules for differentiating inverse functions may be derived by writing the functions in direct form and applying the implicit differentiation technique. For example, $y = \sin^{-1} x$ can be written $x = \sin y$ and differentiating this equation with respect to x gives

$$1 = \cos y \frac{dy}{dx},$$

i.e.

$$\frac{dy}{dx} = \frac{1}{\cos y}$$

$$= \frac{1}{\sqrt{1 - \sin^2 y}}$$

$$= \frac{1}{\sqrt{1 - x^2}}.$$

We take the *positive* square root since $\sin^{-1} x$ is an *increasing* function of x (see Fig. 9.1).

Therefore, we have the rule

$$\frac{d}{dx}(\sin^{-1} x) = \frac{1}{\sqrt{1 - x^2}} \tag{9.4}$$

and the extended form

$$\frac{d}{dx}(\sin^{-1} u) = \frac{1}{\sqrt{1 - u^2}} \frac{du}{dx}. \tag{9.4a}$$

The following derivatives may be established in a similar way:

$$\frac{\mathrm{d}}{\mathrm{d}x}(\cos^{-1} x) = \frac{-1}{\sqrt{1-x^2}}, \tag{9.5}$$

$$\frac{\mathrm{d}}{\mathrm{d}x}(\tan^{-1} x) = \frac{1}{1+x^2}, \tag{9.6}$$

$$\frac{\mathrm{d}}{\mathrm{d}x}(\sinh^{-1} x) = \frac{1}{\sqrt{x^2+1}}, \tag{9.7}$$

$$\frac{\mathrm{d}}{\mathrm{d}x}(\cosh^{-1} x) = \frac{1}{\sqrt{x^2-1}}, \tag{9.8}$$

$$\frac{\mathrm{d}}{\mathrm{d}x}(\tanh^{-1} x) = \frac{1}{1-x^2}. \tag{9.9}$$

Reversing the above standard forms leads to new integrals. However, it is useful to remember them in the following more general form, where a is a positive constant:

$$\int \frac{\mathrm{d}x}{\sqrt{a^2-x^2}} = \sin^{-1}\left(\frac{x}{a}\right), \tag{9.10}$$

$$\int \frac{\mathrm{d}x}{a^2+x^2} = \frac{1}{a}\tan^{-1}\left(\frac{x}{a}\right), \tag{9.11}$$

$$\int \frac{\mathrm{d}x}{\sqrt{x^2+a^2}} = \sinh^{-1}\left(\frac{x}{a}\right), \tag{9.12}$$

$$\int \frac{\mathrm{d}x}{\sqrt{x^2-a^2}} = \cosh^{-1}\left(\frac{x}{a}\right), \tag{9.13}$$

$$\int \frac{\mathrm{d}x}{a^2-x^2} = \frac{1}{a}\tanh^{-1}\left(\frac{x}{a}\right). \tag{9.14}$$

The above standard integrals may be proved by differentiating the right-hand side (see Example 9.4).

Example 9.3

Differentiate the following functions with respect to x.

(i) $\sin^{-1} 5x$.

(ii) $\cos^{-1}\left(\dfrac{1}{x}\right)$.

(iii) $\tan^{-1}\left(\dfrac{1-x}{1+x}\right)$.

(iv) $(1+x^2)\sinh^{-1} x$.

(i) From equation (9.4a),

$$\frac{d}{dx}(\sin^{-1}5x) = \frac{1}{\sqrt{1-(5x)^2}}\frac{d}{dx}(5x)$$
$$= \frac{5}{\sqrt{1-25x^2}}.$$

(ii)
$$\frac{d}{dx}\left\{\cos^{-1}\left(\frac{1}{x}\right)\right\} = \frac{-1}{\sqrt{1-(1/x)^2}}\frac{d}{dx}\left(\frac{1}{x}\right)$$
$$= \frac{-1}{\sqrt{1-(1/x)^2}}\frac{-1}{x^2}$$
$$= \frac{1}{x\sqrt{x^2-1}}.$$

(iii)
$$\frac{d}{dx}\left\{\tan^{-1}\left(\frac{1-x}{1+x}\right)\right\} = \frac{1}{1+\{(1-x)/(1+x)\}^2}\frac{d}{dx}\left(\frac{1-x}{1+x}\right)$$
$$= \frac{(1+x)^2}{(1+x)^2+(1-x)^2}\,\frac{(1+x)(-1)-(1-x)(1)}{(1+x)^2}$$
$$= \frac{-2}{2+2x^2}$$
$$= -\frac{1}{1+x^2}.$$

(iv)
$$\frac{d}{dx}\{(1+x^2)\sinh^{-1}x\} = (1+x^2)\frac{1}{\sqrt{1+x^2}} + 2x\sinh^{-1}x$$
$$= \sqrt{x^2+1} + 2x\sinh^{-1}x.$$

Example 9.4

Carry out the following differentiations and write down the corresponding integral in each case (a is a positive constant).

(i) $\dfrac{d}{dx}\left\{\sin^{-1}\left(\dfrac{x}{a}\right)\right\}$.

(ii) $\dfrac{d}{dx}\left\{\tan^{-1}\left(\dfrac{x}{a}\right)\right\}$.

(i)
$$\frac{\mathrm{d}}{\mathrm{d}x}\left\{\sin^{-1}\left(\frac{x}{a}\right)\right\} = \frac{1}{\sqrt{1-(x/a)^2}}\,\frac{\mathrm{d}}{\mathrm{d}x}\left(\frac{x}{a}\right)$$
$$= \frac{1}{\sqrt{1-(x/a)^2}}\,\frac{1}{a}$$
$$= \frac{1}{\sqrt{a^2-x^2}}.$$

The corresponding integral is

$$\int \frac{\mathrm{d}x}{\sqrt{a^2-x^2}} = \sin^{-1}\left(\frac{x}{a}\right) + C.$$

(ii)
$$\frac{\mathrm{d}}{\mathrm{d}x}\left\{\tan^{-1}\left(\frac{x}{a}\right)\right\} = \frac{1}{1+(x/a)^2}\,\frac{1}{a}$$
$$= \frac{a}{a^2+x^2}.$$

The corresponding integral is

$$\int \frac{a}{a^2+x^2}\,\mathrm{d}x = \tan^{-1}\left(\frac{x}{a}\right) + C$$

or

$$\int \frac{\mathrm{d}x}{a^2+x^2} = \frac{1}{a}\tan^{-1}\left(\frac{x}{a}\right) + C.$$

Example 9.5

Carry out the following integrations.

(i) $\displaystyle\int \frac{\mathrm{d}x}{\sqrt{9-x^2}}$.

(ii) $\displaystyle\int \frac{\mathrm{d}x}{\sqrt{9-25x^2}}$.

(iii) $\displaystyle\int \frac{\mathrm{d}x}{\sqrt{3-2x-x^2}}$.

(iv) $\displaystyle\int \frac{\mathrm{d}x}{\sqrt{25x^2-9}}$.

(v) $\displaystyle\int_{-1}^{1} \frac{\mathrm{d}x}{3+x^2}$.

(vi) $\displaystyle\int_0^1 \frac{dt}{\sqrt{16+9t^2}}$.

(vii) $\displaystyle\int_{-0.5}^{1.5} \frac{dx}{4x^2+4x+17}$.

(viii) $\displaystyle\int_{-1}^{0} \frac{z\,dz}{\sqrt{3-2z-z^2}}$.

(i) $$\int \frac{dx}{\sqrt{9-x^2}} = \int \frac{dx}{\sqrt{3^2-x^2}}$$

$$= \sin^{-1}\left(\frac{x}{3}\right) + C.$$

(ii)

$$\int \frac{dx}{\sqrt{9-25x^2}} = \int \frac{dx}{\sqrt{3^2-(5x)^2}}.$$

Substitute $u = 5x$, $du = 5\,dx$, $dx = \frac{1}{5}\,du$.

$$\int \frac{dx}{\sqrt{9-25x^2}} = \frac{1}{5}\int \frac{du}{\sqrt{3^2-u^2}}$$

$$= \frac{1}{5}\sin^{-1}\left(\frac{u}{3}\right) + C$$

$$= \frac{1}{5}\sin^{-1}\left(\frac{5x}{3}\right) + C.$$

(iii) Completing the square for the quadratic,

$$3-2x-x^2 = 4-(x+1)^2.$$

$$\int \frac{dx}{\sqrt{3-2x-x^2}} = \int \frac{dx}{\sqrt{2^2-(x+1)^2}}.$$

Substitute $u = x+1$, $du = dx$.

$$\int \frac{dx}{\sqrt{3-2x-x^2}} = \int \frac{du}{\sqrt{2^2-u^2}}$$

$$= \sin^{-1}\left(\frac{u}{2}\right) + C$$

$$= \sin^{-1}\left(\frac{x+1}{2}\right) + C.$$

(iv) Substitute $u = 5x$, $\mathrm{d}u = 5\,\mathrm{d}x$ as in (ii).

$$\int \frac{\mathrm{d}x}{\sqrt{25x^2 - 9}} = \tfrac{1}{5}\int \frac{\mathrm{d}u}{\sqrt{u^2 - 3^2}}$$
$$= \tfrac{1}{5}\cosh^{-1}\left(\frac{u}{3}\right) + C$$
$$= \tfrac{1}{5}\cosh^{-1}\left(\frac{5x}{3}\right) + C.$$

(v) $$\int_{-1}^{1} \frac{\mathrm{d}x}{3 + x^2} = \int_{-1}^{1} \frac{\mathrm{d}x}{(\sqrt{3})^2 + x^2}$$
$$= \frac{1}{\sqrt{3}}\left[\tan^{-1}\left(\frac{x}{\sqrt{3}}\right)\right]_{-1}^{1}$$
$$= \frac{1}{\sqrt{3}}\left\{\tan^{-1}\left(\frac{1}{\sqrt{3}}\right) - \tan^{-1}\left(\frac{-1}{\sqrt{3}}\right)\right\}$$
$$= \frac{1}{\sqrt{3}}\left\{\frac{\pi}{6} - \left(-\frac{\pi}{6}\right)\right\}$$
$$= \frac{\pi}{3\sqrt{3}}.$$

(vi) $$\int_{0}^{1} \frac{\mathrm{d}t}{\sqrt{16 + 9t^2}} = \int_{0}^{1} \frac{\mathrm{d}t}{\sqrt{(4)^2 + (3t)^2}}.$$

Substitute $u = 3t$, $\mathrm{d}u = 3\,\mathrm{d}t$, $\mathrm{d}t = \tfrac{1}{3}\,\mathrm{d}u$. When $t = 0, u = 0$; when $t = 1, u = 3$.

$$\int_{0}^{1} \frac{\mathrm{d}t}{\sqrt{16 + 9t^2}} = \tfrac{1}{3}\int_{0}^{3} \frac{\mathrm{d}u}{\sqrt{u^2 + 4^2}}$$
$$= \tfrac{1}{3}\left[\sinh^{-1}\left(\frac{u}{4}\right)\right]_{0}^{3}$$
$$= \tfrac{1}{3}\left\{\sinh^{-1}\left(\tfrac{3}{4}\right) - \sinh^{-1} 0\right\}$$
$$= \tfrac{1}{3}\left\{\ln\left(\tfrac{3}{4} + \sqrt{\tfrac{9}{16} + 1}\right) - 0\right\}$$
$$= \tfrac{1}{3}\ln 2.$$

(vii) Completing the square for the quadratic,

$$4x^2 + 4x + 17 = (2x + 1)^2 + 16.$$

$$\int_{-0.5}^{1.5} \frac{dx}{4x^2+4x+17} = \int_{-0.5}^{1.5} \frac{dx}{(2x+1)^2+4^2}:$$

Substitute $u = 2x + 1$, $du = 2\,dx$, $dx = \frac{1}{2}\,du$. When $x = -0.5$, $u = 0$; when $x = 1.5, u = 4$.

$$\begin{aligned}\int_{-0.5}^{1.5} \frac{dx}{4x^2+4x+17} &= \tfrac{1}{2}\int_0^4 \frac{du}{u^2+4^2}\\ &= \tfrac{1}{2}\left[\tfrac{1}{4}\tan^{-1}\left(\frac{u}{4}\right)\right]_0^4\\ &= \tfrac{1}{8}(\tan^{-1} 1 - \tan^{-1} 0)\\ &= \frac{\pi}{32}.\end{aligned}$$

(viii) $\displaystyle\int_{-1}^{0} \frac{z\,dz}{\sqrt{3-2z-z^2}}$.

Completing the square as in (iii),

$$\int_{-1}^{0} \frac{z\,dz}{\sqrt{3-2z-z^2}} = \int_{-1}^{0} \frac{z\,dz}{\sqrt{2^2-(z+1)^2}}.$$

Substitute $u = z + 1$, $du = dz$; also $z = u - 1$. When $z = -1$, $u = 0$; when $z = 0$, $u = 1$.

$$\int_{-1}^{0} \frac{z\,dz}{\sqrt{3-2z-z^2}} = \int_0^1 \frac{(u-1)\,du}{\sqrt{2^2-u^2}}.$$

The right-hand side can be written as the difference of two integrals:

$$\int_0^1 \frac{u\,du}{\sqrt{2^2-u^2}} - \int_0^1 \frac{du}{\sqrt{2^2-u^2}}.$$

The second of these integrals is a standard form and the first may be completed using the substitution $w = 4 - u^2$, $dw = -2u\,du$. Therefore the required integral is given by

$$\begin{aligned}\int_4^3 \frac{(-\frac{1}{2}\,dw)}{\sqrt{w}} - \int_0^1 \frac{du}{\sqrt{2^2-u^2}} &= [-\sqrt{w}\,]_4^3 - \left[\sin^{-1}\left(\frac{u}{2}\right)\right]_0^1\\ &= -(\sqrt{3}-2) - \{\sin^{-1}(\tfrac{1}{2}) - \sin^{-1} 0\}\\ &= 2 - \sqrt{3} - \frac{\pi}{6}\\ &\approx -0.256.\end{aligned}$$

Example 9.6
Calculate the area enclosed between the curve

$$y = \frac{1}{x^2 + 0.25}$$

and the lines $x = \frac{1}{2}$, $x = \sqrt{3}/2$ and $y = 0$.

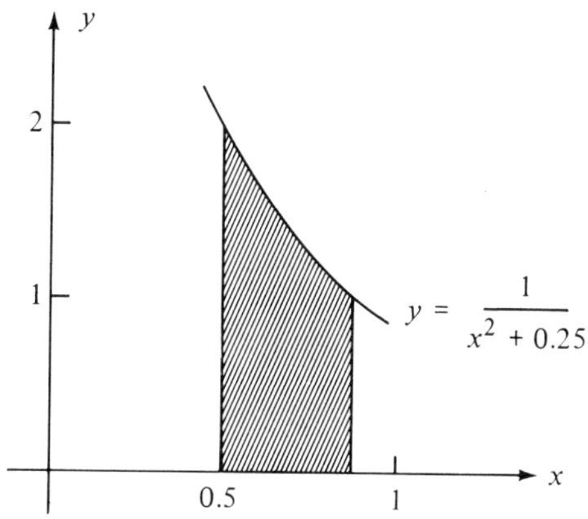

Fig. 9.2

The required area is shown in Fig. 9.2. It is clear that the area is approximately (trapezium approximation) $\frac{1}{2}(2 + 1)(0.866 - 0.5) = 0.549$ units ^{2}.

This calculation will be used to check the exact evaluation of the area by integration:

$$\begin{aligned} \text{area} &= \int_{1/2}^{\sqrt{3}/2} \frac{dx}{(0.5)^2 + x^2} \\ &= \frac{1}{0.5}\left[\tan^{-1}\left(\frac{x}{0.5}\right)\right]_{1/2}^{\sqrt{3}/2} \\ &= 2\left\{\tan^{-1}(\sqrt{3}) - \tan^{-1} 1\right\} \\ &= 2\left(\frac{\pi}{3} - \frac{\pi}{4}\right) \\ &= \frac{\pi}{6} \text{ units}^2 \\ &\approx 0.524, \end{aligned}$$

which is a reasonable comparison with the approximate calculation.

This simple example demonstrates the importance of using *radian measure* when evaluating inverse functions.

Problems

1. Differentiate the following functions with respect to x.
   (i) $\tan^{-1}(7x-1)$.
   (ii) $\sin^{-1}\left(\frac{2}{x}\right)$.
   (iii) $\sqrt{1-x^2}\,\sin^{-1} x$.
   (iv) $\cosh^{-1}\left(\frac{1}{x}\right)$.
   (v) $\tanh^{-1}\left(\frac{4+x}{1+4x}\right)$.
   (vi) $\sinh^{-1}\left(\sqrt{\frac{3}{x}}\right)$.
   (vii) $\frac{1}{(\sin^{-1} x)^2}$.

2. (i) Show that $y=\cosh(2\sin^{-1} t)$ satisfies the equation

$$(1-t^2)\frac{d^2y}{dt^2}-t\frac{dy}{dt}-4y=0.$$

   (ii) Given that $x=\sin(3\tan^{-1}\theta)$ show that

$$(1+\theta^2)^2\frac{d^2x}{d\theta^2}+2\theta(1+\theta^2)\frac{dx}{d\theta}+9x=0.$$

3. Carry out the following integrations.
   (i) $\int\frac{dx}{\sqrt{1-9x^2}}$.
   (ii) $\int\frac{dx}{3+4x^2}$.
   (iii) $\int\frac{dx}{\sqrt{4x^2+3}}$.
   (iv) $\int\frac{dx}{\sqrt{x^2-2x}}$.
   (v) $\int\frac{dx}{x^2+2x+5}$.

(vi) $\displaystyle\int \frac{x\,dx}{x^2 + 4x + 9}$.

4. Evaluate the following.

(i) $\displaystyle\int_0^1 \frac{dx}{\sqrt{4 - x^2}}$.

(ii) $\displaystyle\int_{-\sqrt{3}}^{\sqrt{3}} \frac{du}{9 + u^2}$.

(iii) $\displaystyle\int_0^{1.2} \frac{dx}{36 + 25x^2}$.

(iv) $\displaystyle\int_{-1}^{1} \frac{dt}{\sqrt{9 + 4t^2}}$.

(v) $\displaystyle\int_{1.5}^{2.5} \frac{dz}{\sqrt{4z^2 - 9}}$.

(vi) $\displaystyle\int_{2.0}^{3.5} \frac{dt}{\sqrt{5 + 4t - t^2}}$.

(vii) $\displaystyle\int_{1.0}^{1.5} \frac{3x\,dx}{\sqrt{2x - x^2}}$.

(viii) $\displaystyle\int_{-0.5}^{0} \frac{dx}{1 - x - 2x^2}$.

5. Calculate the area enclosed by the curve

$$y = \frac{1}{\sqrt{1 - 0.5x^2}}$$

and the lines $y = 0$, $x = 0$, and $x = 1$. What is the volume of the solid generated when this area is rotated about the x axis?

Answers

1. (i) $\displaystyle\frac{7}{2 - 14x + 49x^2}$.

(ii) $\displaystyle -\frac{2}{x\sqrt{x^2 - 4}}$.

(iii) $\displaystyle 1 - \frac{x \sin^{-1} x}{\sqrt{1 - x^2}}$.

(iv) $-\dfrac{1}{x\sqrt{1-x^2}}$.

(v) $\dfrac{1}{1-x^2}$.

(vi) $-\dfrac{3}{2x\sqrt{3(3+x)}}$.

(vii) $-\dfrac{2}{\sqrt{1-x^2}\,(\sin^{-1} x)^3}$.

3. (i) $\frac{1}{3}\sin^{-1} 3x + C$.

(ii) $\dfrac{1}{2\sqrt{3}}\tan^{-1}\left(\dfrac{2x}{\sqrt{3}}\right) + C$.

(iii) $\frac{1}{2}\sinh^{-1}\left(\dfrac{2x}{\sqrt{3}}\right) + C$.

(iv) $\cosh^{-1}(x-1) + C$.

(v) $\frac{1}{2}\tan^{-1}\left(\dfrac{x+1}{2}\right) + C$.

(vi) $\frac{1}{2}\ln(x^2 + 4x + 9) - \dfrac{2}{\sqrt{5}}\tan^{-1}\left(\dfrac{x+2}{\sqrt{5}}\right) + C$.

4. (i) $\dfrac{\pi}{6}$.

(ii) $\dfrac{\pi}{9}$.

(iii) $\dfrac{\pi}{120}$.

(iv) $\ln\left(\dfrac{\sqrt{13}+2}{3}\right)$.

(v) $\frac{1}{2}\ln 3$.

(vi) $\dfrac{\pi}{6}$.

(vii) $\dfrac{\pi}{2} + 3 - \dfrac{3\sqrt{3}}{2}$.

(viii) $\frac{1}{3}\ln 4$.

5. $\dfrac{\sqrt{2}\pi}{4}, \dfrac{\sqrt{2}\pi}{2}\ln\left(\dfrac{\sqrt{2}+1}{\sqrt{2}-1}\right)$.

Further Problems for Chapter 9

1. Evaluate $\sin^{-1}(-1)$. Are the following statements true?
   (i) $\sin\{\sin^{-1}(-1)\} = -1$,

   (ii) $\sin^{-1}\left\{\sin\left(\dfrac{3\pi}{2}\right)\right\} = \dfrac{3\pi}{2}$.

2. Determine x such that $\cosh x = 2.6$.

3. Show that for $-1 \leqslant x \leqslant 1$

$$\sin^{-1} x + \cos^{-1} x = \frac{\pi}{2}.$$

4. Given that $f(t) = \sin^{-1} 4t$, evaluate $f'(0.2)$.

5. Differentiate the function $\cosh^{-1}(1 + 2x)$ with respect to x.

6. Evaluate

$$\int_{-1.5}^{1.5} \frac{dx}{\sqrt{3 - x^2}}.$$

7. Evaluate

$$\int_0^2 \frac{dx}{4 + 3x^2}.$$

8. Write the function

$$\int_0^x \frac{dt}{4 - 9t^2}.$$

in logarithmic form.

Answers

1. $-\dfrac{\pi}{2}$.

   (i) Yes.
   (ii) No.

2. $\ln 5$.

4. $\frac{20}{3}$.

5. $\dfrac{1}{\sqrt{x^2+x}}$.

6. $\dfrac{2\pi}{3}$.

7. $\dfrac{\pi}{6\sqrt{3}}$.

8. $\frac{1}{12}\ln\left(\dfrac{2+3x}{2-3x}\right)+C$.

10

Methods of Integration

10.1 INTRODUCTION

The integration techniques studied in Chapter 4 included reduction to standard forms by simple substitutions. In section 10.2, special substitutions related to the inverse trigonometric and hyperbolic functions are considered. In most cases, these have the effect of removing a square root from the integrand (the function to be integrated).

Section 10.3 deals with the integration of *rational functions* (a rational function is defined as the quotient of two polynomials). The method of partial fractions, summarized in section 0.5, is used to express rational functions as the sum of functions which may be integrated either by simple substitution or by using the standard integrals given in Chapter 9.

A method derived from the product rule for differentiation and known as 'integration by parts' is applied in section 10.4 to the integration of products. The method may also be used to determine the integrals of the inverse functions defined in Chapter 9.

10.2 SPECIAL SUBSTITUTIONS

The term $a^2 - x^2$ may be expressed as a perfect square simply by replacing x by $a \sin \theta$ and using the identity $\cos^2 \theta = 1 - \sin^2 \theta$, i.e.

$$\begin{aligned} a^2 - x^2 &= a^2 - a^2 \sin^2 \theta \\ &= a^2(1 - \sin^2 \theta) \\ &= a^2 \cos^2 \theta \\ &= (a \cos \theta)^2. \end{aligned}$$

It follows that the term $\sqrt{a^2 - x^2}$ in an integrand can be simplified to give $a \cos\theta$ by the substitution $x = a \sin\theta$. To demonstrate this procedure we consider the standard integral

$$\int \frac{\mathrm{d}x}{\sqrt{a^2 - x^2}}.$$

Substitute $x = a \sin\theta$, $\mathrm{d}x = a \cos\theta \, \mathrm{d}\theta$, $\sqrt{a^2 - x^2} = a\cos\theta$, and

$$\begin{aligned}\int \frac{\mathrm{d}x}{\sqrt{a^2 - x^2}} &= \int \frac{a \cos\theta \, \mathrm{d}\theta}{a\cos\theta} \\ &= \int \mathrm{d}\theta \\ &= \theta + C \\ &= \sin^{-1}\left(\frac{x}{a}\right) + C,\end{aligned}$$

since $\sin\theta = x/a$. The technique may be applied to simplify non-standard integrals which include the term $\sqrt{a^2 - x^2}$.

For example, to determine the function $\int \sqrt{a^2 - x^2} \, \mathrm{d}x$, we apply the substitution $x = a \sin\theta$ as above and

$$\begin{aligned}\int \sqrt{a^2 - x^2} \, \mathrm{d}x &= \int a\cos\theta \, a\cos\theta \, \mathrm{d}\theta \\ &= \int a^2 \cos^2\theta \, \mathrm{d}\theta \\ &= \frac{a^2}{2}\int (1 + \cos 2\theta) \, \mathrm{d}\theta \\ &= \frac{a^2}{2}(\theta + \tfrac{1}{2}\sin 2\theta) + C \\ &= \frac{a^2}{2}(\theta + \sin\theta\cos\theta) + C \\ &= \frac{a^2}{2}\left\{\sin^{-1}\left(\frac{x}{a}\right) + \frac{x}{a}\frac{\sqrt{a^2 - x^2}}{a}\right\} + C \\ &= \frac{a^2}{2}\sin^{-1}\left(\frac{x}{a}\right) + \frac{x\sqrt{a^2 - x^2}}{2} + C.\end{aligned}$$

In any application involving the integral $\int \sqrt{a^2 - x^2} \, \mathrm{d}x$ the last three steps are unnecessary since the calculation of a *definite* integral may be simplified by changing the limits of integration.

For example, to confirm the formula for the area of a semicircle of radius a, it is necessary to evaluate the integral

$$\int_{-a}^{a} \sqrt{a^2 - x^2}\, dx$$

(see problem 8 in section 5.3).

Substitute $x = a \sin\theta$, $dx = a\cos\theta\, d\theta$, $\sqrt{a^2 - x^2} = a\cos\theta$. When $x = -a$, $\theta = \sin^{-1}(-1) = -\pi/2$; when $x = a$, $\theta = \sin^{-1} 1 = \pi/2$. Therefore,

$$\begin{aligned}\int_{-a}^{a} \sqrt{a^2 - x^2}\, dx &= \int_{-\pi/2}^{\pi/2} a^2 \cos^2\theta\, d\theta \\ &= \frac{a^2}{2}\left[\theta + \tfrac{1}{2}\sin 2\theta\right]_{-\pi/2}^{\pi/2} \\ &= \frac{a^2}{2}\left\{\left(\frac{\pi}{2} + 0\right) - \left(-\frac{\pi}{2} + 0\right)\right\} \\ &= \frac{\pi a^2}{2}.\end{aligned}$$

A list of similar substitutions is given below. Each one is related to a particular trigonometric or hyperbolic identity.

(i) $x = a\sinh\theta$ reduces $\sqrt{a^2 + x^2}$ to $a\cosh\theta$ since $1 + \sinh^2\theta = \cosh^2\theta$.

(ii) $x = a\cosh\theta$ reduces $\sqrt{x^2 - a^2}$ to $a\sinh\theta$ since $\cosh^2\theta - 1 = \sinh^2\theta$.

(iii) (An alternative to (i)) $x = a\tan\theta$ reduces $\sqrt{a^2 + x^2}$ to $a\sec\theta$ since $1 + \tan^2\theta = \sec^2\theta$.

(iv) (An alternative to (ii)) $x = a\sec\theta$ reduces $\sqrt{x^2 - a^2}$ to $a\tan\theta$ since $\sec^2\theta - 1 = \tan^2\theta$.

Example 10.1

Evaluate

$$\int_0^3 \frac{x^n}{\sqrt{16 + x^2}}\, dx$$

for the cases $n = 0, 1, 2$.

The case $n = 0$ is the standard form

$$\int \frac{dx}{\sqrt{16 + x^2}},$$

and the case $n = 1$ can be determined using the simple substitution $u = 16 + x^2$.

However, *all three* cases can be completed using the special substitution $x = 4\sinh\theta$, $dx = 4\cosh\theta\, d\theta$, $\sqrt{16+x^2} = 4\cosh\theta$. When $x = 0$, $\theta = \sinh^{-1} 0 = 0$; when $x = 3$, $\theta = \sinh^{-1}(\frac{3}{4}) = \ln(\frac{3}{4} + \sqrt{\frac{9}{16}+1}) = \ln 2$. For $n = 0$,

$$\begin{aligned}\int_0^3 \frac{dx}{\sqrt{16+x^2}} &= \int_0^{\ln 2} \frac{4\cosh\theta\, d\theta}{4\cosh\theta}\\ &= \int_0^{\ln 2} d\theta\\ &= [\theta]_0^{\ln 2}\\ &= \ln 2.\end{aligned}$$

For $n = 1$,

$$\begin{aligned}\int_0^3 \frac{x\, dx}{\sqrt{16+x^2}} &= \int_0^{\ln 2} \frac{4\sinh\theta\; 4\cosh\theta\, d\theta}{4\cosh\theta}\\ &= \int_0^{\ln 2} 4\sinh\theta\, d\theta\\ &= 4[\cosh\theta]_0^{\ln 2}\\ &= 4\left(\frac{2+2^{-1}}{2} - 1\right)\\ &= 1.\end{aligned}$$

For $n = 2$,

$$\begin{aligned}\int_0^3 \frac{x^2\, dx}{\sqrt{16+x^2}} &= \int_0^{\ln 2} \frac{(4\sinh\theta)^2\; 4\cosh\theta\, d\theta}{4\cosh\theta}\\ &= \int_0^{\ln 2} 16\sinh^2\theta\, d\theta\\ &= \int_0^{\ln 2} 8(\cosh 2\theta - 1)\, d\theta\\ &= [4\sinh 2\theta - 8\theta]_0^{\ln 2}\\ &= \{4\sinh(\ln 4) - 8\ln 2\} - 0\\ &= 2(4 - 4^{-1}) - 8\ln 2\\ &= 7.5 - 8\ln 2.\end{aligned}$$

Example 10.2
Evaluate the following.

(i) $\displaystyle\int_0^3 \frac{dx}{(9+x^2)^2}$.

(ii) $\displaystyle\int_{-1}^0 \sqrt{x^2-2x}\,dx$.

(i) Although there is no square root in the integrand, either of the substitutions $x = 3\sinh\theta$ or $x = 3\tan\theta$ will reduce $9 + x^2$ to a single term and simplify the integration. In particular, $x = 3\tan\theta$, $dx = 3\sec^2\theta\,d\theta$, $9 + x^2 = 9 + 9\tan^2\theta = 9(1+\tan^2\theta) = 9\sec^2\theta$. When $x = 0$, $\theta = 0$; when $x = 3$, $\theta = \pi/4$. Therefore,

$$\int_0^3 \frac{dx}{(9+x^2)^2} = \int_0^{\pi/4} \frac{3\sec^2\theta\,d\theta}{81\sec^4\theta}$$

$$= \tfrac{1}{27}\int_0^{\pi/4} \cos^2\theta\,d\theta$$

$$= \tfrac{1}{54}\int_0^{\pi/4} (1+\cos 2\theta)\,d\theta$$

$$= \tfrac{1}{54}\left[\theta + \tfrac{1}{2}\sin 2\theta\right]_0^{\pi/4}$$

$$= \frac{\pi+2}{216}.$$

(ii) Completing the square for the quadratic gives

$$\int_{-1}^0 \sqrt{x^2-2x}\,dx = \int_{-1}^0 \sqrt{(x-1)^2-1}\,dx,$$

which is simplified by the successive substitutions $u = 1 - x$, $u = \cosh\theta$; or, alternatively, combining these two substitutions into one, we proceed as follows. Let $1 - x = \cosh\theta$, $-dx = \sinh\theta\,d\theta$, $dx = -\sinh\theta\,d\theta$. $(1-x)^2 - 1 = \cosh^2\theta - 1 = \sinh^2\theta$. When $x = -1$, $\theta = \cosh^{-1}2 = \ln(2+\sqrt{4-1}) = \ln(2+\sqrt{3})$; when $x = 0$, $\theta = \cosh^{-1}1 = 0$. (Note that we chose $1 - x = \cosh\theta$ instead of $x - 1 = \cosh\theta$ so that $\cosh\theta \geqslant 1$ for the interval of integration $-1 \leqslant x \leqslant 0$.)

$$\int_{-1}^0 \sqrt{(x-1)^2-1}\,dx = -\int_{\ln(2+\sqrt{3})}^0 \sinh\theta\sinh\theta\,d\theta$$

$$= \tfrac{1}{2}\int_{\ln(2+\sqrt{3})}^0 (1-\cosh 2\theta)\,d\theta$$

$$= \tfrac{1}{2}\left[\theta - \tfrac{1}{2}\sinh 2\theta\right]_{\ln(2+\sqrt{3})}^0$$

$$= 0 - \tfrac{1}{2}\left\{\ln(2+\sqrt{3}) - \tfrac{1}{4}\left\{(2+\sqrt{3})^2 - (2+\sqrt{3})^{-2}\right)\right\}$$
$$= \sqrt{3} - \tfrac{1}{2}\ln(2+\sqrt{3}).$$

Problems

1. Evaluate the following integrals.

   (i) $\displaystyle\int_0^2 \sqrt{4-x^2}\,\mathrm{d}x.$

   (ii) $\displaystyle\int_0^3 \sqrt{16+x^2}\,\mathrm{d}x.$

   (iii) $\displaystyle\int_{2/3}^{10/9} \sqrt{9u^2-4}\,\mathrm{d}u.$

   (iv) $\displaystyle\int_0^3 x\sqrt{16+x^2}\,\mathrm{d}x.$

   (v) $\displaystyle\int_{-1}^1 \sqrt{3+2x-x^2}\,\mathrm{d}x.$

2. Determine the following functions.

   (i) $\displaystyle\int \sqrt{9+2x^2}\,\mathrm{d}x.$

   (ii) $\displaystyle\int \frac{\mathrm{d}t}{(a^2+t^2)^{3/2}}.$

3. Simplify the integral

$$\int \frac{\mathrm{d}x}{(x^2-1)^{3/2}}$$

   by the following methods.

   (i) The substitution $x = \cosh\theta$.

   (ii) The substitution $x = \sec\theta$.

   Deduce that

$$\int_{\ln(1+\sqrt{2})}^{\ln(2+\sqrt{3})} \operatorname{cosech}^2\theta\,\mathrm{d}\theta = \int_{\pi/4}^{\pi/3} \frac{\cos\theta\,\mathrm{d}\theta}{\sin^2\theta}.$$

   Determine the function

$$\int \frac{\mathrm{d}x}{(x^2-1)^{3/2}}.$$

4. (i) Use the substitution $y = 2\sin^2 x$ to show that

$$\int \sqrt{\frac{y}{2-y}}\,\mathrm{d}y = -\sqrt{y(2-y)} + 2\sin^{-1}\left(\sqrt{\frac{y}{2}}\right) + C.$$

(ii) Carry out the integration

$$\int \sqrt{\frac{y}{y-2}}\,\mathrm{d}y.$$

Answers

1. (i) π.
   (ii) $\frac{15}{2} + 8\ln 2$.
   (iii) $\frac{2}{3}(\frac{20}{9} - \ln 3)$.
   (iv) $\frac{61}{3}$.
   (v) π.

2. (i) $\dfrac{9}{2\sqrt{2}}\sinh^{-1}\left(\dfrac{\sqrt{2}x}{3}\right) + \dfrac{x}{2}\sqrt{9+2x^2} + C.$

   (ii) $\dfrac{t}{a^2\sqrt{a^2+t^2}} + C.$

3. $-\dfrac{x}{\sqrt{1-x^2}} + C.$

4. $2\cosh^{-1}\left(\sqrt{\dfrac{y}{2}}\right) + \sqrt{y(y-2)} + C.$

10.3 INTEGRATION OF RATIONAL FUNCTIONS

A rational function may be integrated by expressing the function in partial fraction form according to the procedure summarized in section 0.5. This will lead to a number of simpler integrals which may include the following types.

(i) $\displaystyle\int \frac{A\,\mathrm{d}x}{(kx+l)^n}$, n a positive integer.

(ii) $\displaystyle\int \frac{(Bx+C)\,\mathrm{d}x}{ax^2+bx+c}$, $b^2 < 4ac$.

Type (i) is reduced to

$$\int \frac{A}{k}u^{-n}\,\mathrm{d}u$$

by the substitution $u = kx + l$. Type (ii) was studied in Chapter 9 and requires the technique of completing the square.

A rational function of $\sin x$ and/or $\cos x$ may be reduced to a rational function of the variable t by the substitution

$$t = \tan\left(\frac{x}{2}\right),$$

$$\begin{aligned} dt &= \tfrac{1}{2}\sec^2\left(\frac{x}{2}\right) dx \\ &= \tfrac{1}{2}\left\{1 + \tan^2\left(\frac{x}{2}\right)\right\} dx \\ &= \tfrac{1}{2}(1 + t^2)\, dx, \end{aligned}$$

i.e.

$$dx = \frac{2\, dt}{1 + t^2}.$$

$$\begin{aligned} \sin x &= 2\sin\left(\frac{x}{2}\right)\cos\left(\frac{x}{2}\right) \\ &= \frac{2\tan(x/2)}{\sec^2(x/2)} \\ &= \frac{2\tan(x/2)}{1 + \tan^2(x/2)} \\ &= \frac{2t}{1 + t^2}. \end{aligned}$$

$$\begin{aligned} \cos x &= \cos^2\left(\frac{x}{2}\right) - \sin^2\left(\frac{x}{2}\right) \\ &= \frac{1 - \tan^2(x/2)}{\sec^2(x/2)} \\ &= \frac{1 - \tan^2(x/2)}{1 + \tan^2(x/2)} \\ &= \frac{1 - t^2}{1 + t^2}. \end{aligned}$$

Example 10.5 demonstrates the application of this substitution.

Example 10.3
Carry out the following integrations.

(i) $$\int \frac{x^2 + x + 8}{(x+1)(x^2 + 2x + 5)}\,\mathrm{d}x.$$

(ii) $$\int \frac{11x - 3}{(x-1)(3x-1)^2}\,\mathrm{d}x.$$

(i) Expressing the integrand in partial fraction form gives

$$\frac{x^2 + x + 8}{(x+1)(x^2 + 2x + 5)} = \frac{2}{x+1} - \frac{x+2}{x^2 + 2x + 5}.$$

Therefore,

$$\int \frac{x^2 + x + 8}{(x+1)(x^2 + 2x + 5)}\,\mathrm{d}x = \int \frac{2}{x+1}\,\mathrm{d}x - \int \frac{x+2}{x^2 + 2x + 5}\,\mathrm{d}x$$

$$= 2\ln(x+1) - \int \frac{x+2}{(x+1)^2 + 2^2}\,\mathrm{d}x$$

If we put $u = x + 1$, in the integral, then

$$2\ln(x+1) - \int \frac{x+2}{(x+1)^2 + 2^2}\,\mathrm{d}x = 2\ln(x+1) - \int \frac{(u+1)\,\mathrm{d}u}{u^2 + 2^2}$$

$$= 2\ln(x+1) - \int \frac{u\,\mathrm{d}u}{u^2 + 2^2} - \int \frac{\mathrm{d}u}{u^2 + 2^2}$$

$$= 2\ln(x+1) - \tfrac{1}{2}\ln(u^2 + 4) - \tfrac{1}{2}\tan^{-1}\left(\frac{u}{2}\right) + C$$

$$= 2\ln(x+1) - \tfrac{1}{2}\ln(x^2 + 2x + 5) - \tfrac{1}{2}\tan^{-1}\left(\frac{x+1}{2}\right) + C.$$

(ii) Expressing the integrand in partial fraction form gives

$$\frac{11x - 3}{(x-1)(3x-1)^2} = \frac{2}{x-1} - \frac{6}{3x-1} - \frac{1}{(3x-1)^2}.$$

Therefore,

$$\int \frac{11x - 3}{(x-1)(3x-1)^2}\,\mathrm{d}x = \int \frac{2\,\mathrm{d}x}{x-1} - \int \frac{6\,\mathrm{d}x}{3x-1} - \int \frac{\mathrm{d}x}{(3x-1)^2}$$

$$= 2\ln(x-1) - \int \frac{6\,\mathrm{d}x}{3x-1} - \int \frac{\mathrm{d}x}{(3x-1)^2}$$

If we put $u = 3x - 1$ in the integrals, then

$$2\ln(x-1) - \int \frac{6\,\mathrm{d}x}{3x-1} - \int \frac{\mathrm{d}x}{(3x-1)^2} = 2\ln(x-1) - 2\int \frac{\mathrm{d}u}{u} - \tfrac{1}{3}\int \frac{\mathrm{d}u}{u^2}$$

$$= 2\ln(x-1) - 2\ln u + \frac{1}{3u} + C$$

$$= 2\ln(x-1) - 2\ln(3x-1) + \frac{1}{3(3x-1)} + C.$$

Example 10.4
Evaluate

$$\int_0^1 \frac{2x^3 + 9x^2 + 14x + 6}{2x^2 + 3x + 1}\,\mathrm{d}x.$$

The degree of the numerator is not less than the degree of the denominator. Therefore a long division must be done, giving the quotient $x + 3$ and remainder $4x + 3$.

$$\frac{2x^3 + 9x^2 + 14x + 6}{2x^2 + 3x + 1} = x + 3 + \frac{4x+3}{2x^2 + 3x + 1}$$

$$= x + 3 + \frac{4x+3}{(x+1)(2x+1)}$$

$$= x + 3 + \frac{1}{x+1} + \frac{2}{2x+1}.$$

$$\int_0^1 \frac{2x^3 + 9x^2 + 14x + 6}{2x^2 + 3x + 1}\,\mathrm{d}x = \int_0^1 (x+3)\,\mathrm{d}x + \int_0^1 \frac{\mathrm{d}x}{x+1} + \int_0^1 \frac{2\,\mathrm{d}x}{2x+1}$$

$$= \left[\frac{x^2}{2} + 3x + \ln(x+1) + \ln(2x+1)\right]_0^1$$

$$= (\tfrac{1}{2} + 3 + \ln 2 + \ln 3) - (\ln 1 + \ln 1)$$

$$= \tfrac{7}{2} + \ln 6.$$

Example 10.5
Use the substitution

$$t = \tan\left(\frac{x}{2}\right)$$

to evaluate

$$\int_0^{\pi/2} \frac{dx}{12 + 13 \cos x + 12 \sin x} .$$

Let

$$t = \tan\left(\frac{x}{2}\right),$$

$$dx = \frac{2\, dt}{1 + t^2},$$

$$\sin x = \frac{2t}{1 + t^2},$$

$$\cos x = \frac{1 - t^2}{1 + t^2} .$$

When $x = 0, t = 0$; when $x = \pi/2, t = 1$. Therefore,

$$\int_0^{\pi/2} \frac{dx}{12 + 13 \cos x + 12 \sin x}$$

$$= \int_0^1 \frac{2\, dt/(1 + t^2)}{12 + 13\{(1 - t^2)/(1 + t^2)\} + 12\{2t/(1 + t^2)\}}$$

$$\int_0^1 \frac{2\, dt}{12(1 + t^2) + 13(1 - t^2) + 12(2t)}$$

$$= \int_0^1 \frac{2\, dt}{25 + 24t - t^2} .$$

The substitution

$$t = \tan\left(\frac{x}{2}\right)$$

has reduced the integral to a rational function of t. We factorize the quadratic denominator and express the rational function in partial fraction form:

$$\frac{2}{25 + 24t - t^2} = \frac{2}{(25 - t)(1 + t)}$$

$$= \tfrac{1}{13}\left(\frac{1}{25 - t} + \frac{1}{1 + t}\right) .$$

Therefore the required integral is

$$\tfrac{1}{13}\int_0^1 \left(\frac{1}{25-t}+\frac{1}{1+t}\right)\mathrm{d}t = \tfrac{1}{13}\left[-\ln(25-t)+\ln(1+t)\right]_0^1$$
$$= \tfrac{1}{13}\ln\left(\tfrac{25}{12}\right).$$

Problems

1. Carry out the following integrations.

(i) $\displaystyle\int \frac{2x+15}{x^2+5x+6}\,\mathrm{d}x.$

(ii) $\displaystyle\int \frac{x^2-2x-6}{x^2-x-2}\,\mathrm{d}x.$

(iii) $\displaystyle\int \frac{x^2+5}{(x+2)(2x-1)^2}\,\mathrm{d}x.$

(iv) $\displaystyle\int \frac{x^2+4x-7}{(x+1)(x^2+4)}\,\mathrm{d}x.$

(v) $\displaystyle\int \frac{4x^2+5x+5}{x^3+2x^2+5x}\,\mathrm{d}x.$

(vi) $\displaystyle\int \frac{x^4}{x^3+8}\,\mathrm{d}x.$

2. Evaluate the following.

(i) $\displaystyle\int_0^2 \frac{\mathrm{d}t}{(t+1)^2(t^2+4)}.$

(ii) $\displaystyle\int_0^2 \frac{x\,\mathrm{d}x}{(x^2+1)(4x^2+1)}.$

(Hint: put $u=x^2$.)

(iii) $\displaystyle\int_0^1 \frac{x+3}{(x+1)^4}\,\mathrm{d}x.$

(Hint: partial fractions are unnecessary.)

3. Use the substitution

$$t=\tan\left(\frac{x}{2}\right)$$

to obtain the following integrals.

(i) $\int \frac{dx}{\sin x}$.

(ii) $\int \frac{dx}{\cos x}$.

(iii) $\int_0^{\pi/2} \frac{dx}{2 + \sin x}$.

Answers

1. (i) $11 \ln(x+2) - 9 \ln(x+3) + C$.

(ii) $x + \ln(x+1) - 2\ln(x-2) + C$.

(iii) $\frac{9}{25} \ln(x+2) - \frac{11}{100} \ln(2x-1) - \frac{21}{20}(2x-1)^{-1} + C$.

(iv) $-2\ln(x+1) + \frac{3}{2}\ln(x^2+4) + \frac{1}{2}\tan^{-1}\left(\frac{x}{2}\right) + C$.

(v) $\ln x + \frac{3}{2}\ln(x^2+2x+5) + C$.

(vi) $\frac{x^2}{2} + \frac{4}{3}\ln(x+2) - \frac{2}{3}\ln(x^2-2x+4) - \frac{4}{\sqrt{3}}\tan^{-1}\left(\frac{x-1}{\sqrt{3}}\right) + C$.

2. (i) $\frac{1}{25}\ln(\frac{9}{2}) + \frac{1}{15} - \frac{3\pi}{200}$.

(ii) $\frac{1}{6}\ln(\frac{17}{5})$.

(iii) $\frac{23}{24}$.

3. (i) $\ln\left\{\tan\left(\frac{x}{2}\right)\right\} + C$.

(ii) $\ln\left\{\frac{1+\tan(x/2)}{1-\tan(x/2)}\right\} + C$.

(iii) $\frac{\pi}{3\sqrt{3}}$.

10.4 INTEGRATION BY PARTS

We recall the product rule for differentiation, equation (2.1a):

$$\frac{d}{dx}(uv) = u\frac{dv}{dx} + v\frac{du}{dx}$$

where u and v are functions of x.

Integrating equation (2.1a) gives

$$uv = \int u \frac{dv}{dx}\,dx + \int v \frac{du}{dx}\,dx$$

i.e.

$$uv = u\,dv + \int v\,du$$

or

$$\int u\,dv = uv - \int v\,du, \tag{10.1}$$

where du and dv are the differentials of u and v as defined in section 4.2.

Equation (10.1) is known as the rule for **integration by parts**. It is useful when integrating the product of two functions, $\int f(x)g(x)\,dx$, and for integrating inverse functions.

When carrying out the integration $\int f(x)g(x)\,dx$, it is first necessary to *define* u and dv, and then to *determine* du and v. The choice of u and dv satisfying $u\,dv = f(x)g(x)\,dx$ is not unique, but the decision in making this choice normally requires that the integral $\int v\,du$ on the right-hand side of equation (10.1) is simpler than the given integral.

For example, to determine the function $\int x\,e^x\,dx$, we choose

$$u = x, \quad dv = e^x\,dx,$$

(note that dx must be included in the definition of the differential dv). Therefore,

$$du = dx, \quad v = \int dv = e^x$$

(constant of integration unnecessary). Applying equation (10.1),

$$\begin{aligned}\int x\,e^x\,dx &= x\,e^x - \int e^x\,dx \\ &= x\,e^x - e^x + C.\end{aligned}$$

When applied to definite integrals, equation (10.1) can be written

$$\int_a^b u\,dv = [u\,v]_a^b - \int_a^b v\,du, \tag{10.2}$$

where a and b refer to values of the independent variable associated with the functions u and v.

For example,

$$\begin{aligned}\int_0^1 x\,e^x\,dx &= [x\,e^x]_0^1 - \int_0^1 e^x\,dx \\ &= e - [e^x]_0^1 \\ &= e - (e - 1) \\ &= 1.\end{aligned}$$

Example 10.6
Carry out the following integrations.

(i) $\int x \cos 2x \, \mathrm{d}x$.

(ii) $\int x^2 \, \mathrm{e}^x \, \mathrm{d}x$.

(iii) $\int x \ln x \, \mathrm{d}x$.

(i) We choose $u = x$, $\mathrm{d}v = \cos 2x \, \mathrm{d}x$. Therefore, $\mathrm{d}u = \mathrm{d}x$, $v = \int \cos 2x \, \mathrm{d}x = \frac{1}{2} \sin 2x$. Applying equation (10.1),

$$\int x \cos 2x \, \mathrm{d}x = x \tfrac{1}{2} \sin 2x - \int \tfrac{1}{2} \sin 2x \, \mathrm{d}x$$

$$= \tfrac{1}{2} x \sin 2x + \tfrac{1}{4} \cos 2x + C.$$

(ii) Let $u = x^2$, $\mathrm{d}v = \mathrm{e}^x \, \mathrm{d}x$. Therefore $\mathrm{d}u = 2x \, \mathrm{d}x$, $v = \mathrm{e}^x$, and

$$\int x^2 \, \mathrm{e}^x \, \mathrm{d}x = x^2 \, \mathrm{e}^x - \int \mathrm{e}^x \, 2x \, \mathrm{d}x$$

$$= x^2 \, \mathrm{e}^x - 2 \int x \, \mathrm{e}^x \, \mathrm{d}x.$$

The integral $x \, \mathrm{e}^x \, \mathrm{d}x$ is simpler than the given integral but requires a second application of the rule for integration by parts. However,

$$\int x \, \mathrm{e}^x \, \mathrm{d}x = x \, \mathrm{e}^x - \mathrm{e}^x + C$$

(see above). Therefore,

$$\int x^2 \, \mathrm{e}^x \, \mathrm{d}x = x^2 \, \mathrm{e}^x - 2x \, \mathrm{e}^x + 2 \, \mathrm{e}^x + C.$$

(iii) To apply equation (10.1) to $\int x \ln \mathrm{d}x$ the 'obvious' choice for u and $\mathrm{d}v$, based on previous examples, is $u = x$, $\mathrm{d}v = \ln x \, \mathrm{d}x$, but this requires $v = \int \ln x \, \mathrm{d}x$ which is not one of the standard integrals. Since we do know how to *differentiate* the function $\ln x$, we choose

$$u = \ln x, \qquad \mathrm{d}v = x \, \mathrm{d}x.$$

Therefore,

$$\mathrm{d}u = \frac{1}{x} \, \mathrm{d}x, \qquad v = \frac{x^2}{2},$$

and

$$\int x \ln x \, dx = (\ln x)\frac{x^2}{2} - \int \frac{x^2}{2}\,\frac{1}{x}\, dx$$

$$= \frac{x^2}{2}\ln x - \int \frac{x}{2}\, dx$$

$$= \frac{x^2}{2}\ln x - \frac{x^2}{4} + C.$$

Example 10.7

Apply the method of integration by parts to the following integrals.

(i) $\int \ln x \, dx$.

(ii) $\int \sin^{-1} x \, dx$.

(i) Following the reasoning of Example 10.6 (iii), we choose

$$u = \ln x, \qquad dv = dx.$$

Therefore,

$$du = \frac{1}{x}\, dx, \qquad v = x$$

and

$$\int \ln x \, dx = (\ln x)x - \int x \frac{1}{x}\, dx$$

$$= x \ln x - \int dx$$

$$= x \ln x - x + C.$$

(ii) Let $u = \sin^{-1} x$, $dv = dx$. Therefore,

$$du = \frac{dx}{\sqrt{1-x^2}}, \qquad v = x$$

and

$$\int \sin^{-1} x \, dx = (\sin^{-1} x)x - \int x \frac{1}{\sqrt{1-x^2}}\, dx.$$

Now substitute $w = 1 - x^2$, $dw = -2x \, dx$, $x \, dx = -\frac{1}{2}\, dw$.

$$\int \sin^{-1} x \, dx = x \sin^{-1} x - \int w^{-1/2} \left(-\tfrac{1}{2} \, dw\right)$$
$$= x \sin^{-1} x + \tfrac{1}{2} 2w^{1/2} + C$$
$$= x \sin^{-1} x + \sqrt{1-x^2} + C.$$

Example 10.8

Evaluate the following.

(i) $\displaystyle\int_0^{\pi/2} e^{2x} \cos x \, dx.$

(ii) $\displaystyle\int_0^{\pi/2} e^{2x} \sin x \, dx.$

(i) Let $u = e^{2x}$, $dv = \cos x \, dx$. Therefore, $du = 2\, e^{2x} \, dx$, $v = \sin x$. Applying equation (10.2),

$$\int_0^{\pi/2} e^{2x} \cos x \, dx = [e^{2x} \sin x]_0^{\pi/2} - 2 \int_0^{\pi/2} e^{2x} \sin x \, dx$$
$$= e^{\pi} - 2 \int_0^{\pi/2} e^{2x} \sin x \, dx.$$

Denote

$$\int_0^{\pi/2} e^{2x} \cos x \, dx$$

by I_C and

$$\int_0^{\pi/2} e^{2x} \sin x \, dx$$

by I_S. Therefore,

$$I_C = e^{\pi} - 2I_S. \qquad (10.3)$$

To evaluate I_S let $u = e^{2x}$, $dv = \sin x \, dx$. Therefore, $du = 2e^{2x} \, dx$, $v = -\cos x$ and

$$I_S = [-e^{2x} \cos x]_0^{\pi/2} - (-2) \int_0^{\pi/2} e^{2x} \cos x \, dx.$$

Note that the integral on the right-hand side is in fact the given integral I_C. This allows the evaluation to be completed since

$$I_S = 1 + 2I_C \qquad (10.4)$$

and equations (10.3) and (10.4) may be solved for I_C.

$$I_C = e^{\pi} - 2(1 + 2I_C)$$

i.e.

$$I_C = \tfrac{1}{5}(e^{\pi} - 2).$$

(ii) To evaluate

$$\int_0^{\pi/2} e^{2x} \sin x \, dx,$$

we simply solve equations (10.3) and (10.4) for I_S:

$$I_S = 1 + \tfrac{2}{5}(e^{\pi} - 2)$$
$$= \tfrac{1}{5}(1 + 2\,e^{\pi}).$$

Example 10.9
Given that

$$C(n) = \int_0^{\pi/2} \cos^n x \, dx,$$

show that

$$C(n) = \frac{n-1}{n} \; C(n-2).$$

Hence evaluate the following.

(i) $\int_0^{\pi/2} \cos^6 x \, dx.$

(ii) $\int_0^{\pi/2} \cos^7 x \, dx.$

$$C(n) = \int_0^{\pi/2} \cos^{n-1} x \cos x \, dx.$$

Let $u = \cos^{n-1} x$, $dv = \cos x \, dx$. Therefore, $du = -(n-1) \cos^{n-2} x \sin x \, dx$, $v = \sin x$ and

$$C(n) = [\cos^{n-1} x \sin x]_0^{\pi/2} + (n-1) \int_0^{\pi/2} \cos^{n-2} x \sin^2 x \, dx$$

$$= 0 + (n-1) \int_0^{\pi/2} \cos^{n-2} x(1 - \cos^2 x) \, dx$$

$$= (n-1) \int_0^{\pi/2} (\cos^{n-2} x - \cos^n x) \, dx$$

i.e.

$$C(n) = (n-1)\{C(n-2) - C(n)\}.$$

Solving for $C(n)$,

$$C(n) = \frac{n-1}{n}\, C(n-2). \tag{10.5}$$

Equation (10.5) is known as a **reduction formula**; it can be applied successively until $C(n)$ is expressed as a multiple of either $C(1)$ or $C(0)$.

$$C(1) = \int_0^{\pi/2} \cos x \, dx = 1,$$

$$C(0) = \int_0^{\pi/2} 1 \, dx = \frac{\pi}{2}.$$

For example,

(i)
$$\begin{aligned}\int_0^{\pi/2} \cos^6 x \, dx &= C(6)\\ &= \tfrac{5}{6}\, C(4)\\ &= \tfrac{5}{6}\tfrac{3}{4}\, C(2)\\ &= \tfrac{5}{6}\tfrac{3}{4}\tfrac{1}{2}\, C(0)\\ &= \frac{5}{6}\frac{3}{4}\frac{1}{2}\frac{\pi}{2}\\ &= \frac{5\pi}{32}.\end{aligned}$$

(ii)
$$\begin{aligned}\int_0^{\pi/2} \cos^7 x \, dx &= C(7)\\ &= \tfrac{6}{7}\, C(5)\\ &= \tfrac{6}{7}\tfrac{4}{5}\, C(3)\\ &= \tfrac{6}{7}\tfrac{4}{5}\tfrac{2}{3}\, C(1)\\ &= \tfrac{6}{7}\tfrac{4}{5}\tfrac{2}{3}\, 1\\ &= \tfrac{16}{35}.\end{aligned}$$

Problems

1. Carry out the following integrations.

   (i) $\int x \sin 3x \, dx$.

   (ii) $\int x^2 \cos x \, dx$.

   (iii) $\int \tan^{-1} x \, dx$.

(iv) $\int \sinh^{-1} x \, dx$.

(v) $\int e^x \sin x \, dx$.

2. Evaluate the following.

(i) $\int_0^{\pi/2} x \sin x \, dx$.

(ii) $\int_1^2 \cosh^{-1} u \, du$.

3. Show that, for $n \neq -1$,

$$\int x^n \ln x \, dx = \frac{x^{n+1}}{n+1}\left(\ln x - \frac{1}{n+1}\right) + C.$$

What is the result when $n = -1$?

4. Using the technique demonstrated in Example 10.9 show that

$$S(n) = \frac{n-1}{n} S(n-2),$$

where $S(n)$ is defined by

$$S(n) = \int_0^{\pi/2} \sin^n \theta \, d\theta.$$

Hence write down the value of the following.

(i) $\int_0^{\pi/2} \sin^5 \theta \, d\theta$.

(ii) $\int_0^{\pi/2} \sin^4 t \, dt$.

Show that

$$\int_0^2 (4-x^2)^{3/2} \, dx = 16 \int_0^{\pi/2} \cos^4 \theta \, d\theta = 3\pi.$$

5. The definite integral

$$I_n = \int_0^\infty x^n e^{-x} \, dx$$

is defined as

$$\lim_{R\to\infty} \int_0^R x^n \, e^{-x} \, dx \quad .$$

Given that $R^n \, e^{-R} \to 0$ as $R \to \infty$ show that

$$I_n = nI_{n-1} \, .$$

Deduce that $I_n = n!$ when n is a positive integer.

Answers

1. (i) $-\frac{1}{3} x \cos 3x + \frac{1}{9} \sin 3x + C.$
   (ii) $x^2 \sin x + 2x \cos x - 2 \sin x + C.$
   (iii) $x \tan^{-1} x - \frac{1}{2} \ln(1 + x^2) + C.$
   (iv) $x \sinh^{-1} x - \sqrt{1 + x^2} + C.$
   (v) $\frac{1}{2} e^x (\sin x - \cos x) + C.$

2. (i) 1.
   (ii) $2 \ln(2 + \sqrt{3}) - \sqrt{3}.$

3. $\frac{1}{2} (\ln x)^2 + C.$

4. (i) $\frac{8}{15}$.

   (ii) $\dfrac{3\pi}{16}$.

Further Problems for Chapter 10

1. What substitution would you use to simplify the integral $\int (x^2 - 36)^{5/2} \, dx$?

2. Without evaluating the integrals, show that

$$\int_0^a (a^2 + x^2) \, dx = a^3 \int_0^{\pi/4} \sec^4 \theta \, d\theta.$$

3. Applying the substitution $x = a \sinh \theta$ to the integral in problem 2, show that

$$\int_0^{\ln(1+\sqrt{2})} \cosh^3 \theta \, d\theta = \frac{4}{3}.$$

4. By expressing the integrand as the sum of partial fractions, confirm the standard form

$$\int \frac{dx}{a^2 - x^2} = \frac{1}{2a} \ln\left(\frac{a + x}{a - x}\right) + C.$$

5. Determine the function

$$\int \frac{\mathrm{d}x}{x^2 - x}.$$

6. Evaluate

$$\int_0^{\pi/2} x \cos x \,\mathrm{d}x.$$

7. Evaluate

$$\int_0^{\pi/2} \cos^2 x \,\mathrm{d}x$$

using equation (10.5).

8. Given that $I_n = \tan^n x \,\mathrm{d}x$, show that

$$I_n = \frac{\tan^{n-1} x}{n-1} - I_{n-2}, \qquad n \neq 1.$$

(Hint: integration by parts is *not* required; use $\tan^2 x = \sec^2 x - 1$ followed by simple substitution.)

Answers

1. $x = 6 \cosh \theta$ or $x = 6 \sec \theta$.

5. $\ln\left(\frac{x-1}{x}\right) + C.$

6. $\frac{\pi}{2} - 1.$

7. $\frac{\pi}{4}.$

11

Further Applications of Integration

11.1 INTRODUCTION

In Chapter 5, integration was applied to the calculation of areas and volumes. A number of additional applications are considered in this chapter and in each case Theorem 5.1 (fundamental theorem of integral calculus) is applied. As in Chapter 5, the formal procedure of the theorem will be relaxed where appropriate by writing a differential instead of Δx for the strip width and using the integral sign to denote summation.

11.2 THE MEAN VALUE OF A FUNCTION

Consider the function $y = f(x)$ in the interval $a \leqslant x \leqslant b$. Divide the interval into n sub-intervals of width Δx, i.e. $n\,\Delta x = b - a$. The ordinates $y_0, y_1, y_2, \ldots, y_n$ are as shown in Fig. 11.1.

The average value of y in $a \leqslant x \leqslant b$ is approximately

$$\frac{y_0 + y_1 + y_2 + \ldots + y_n}{n+1} = \frac{\sum_{i=0}^{n} y_i\,\Delta x}{(b-a) + \Delta x},$$

since

$$n = \frac{b-a}{\Delta x}.$$

We define the **mean value** of y, denoted by $\bar{y}$, as the limit of this summation as $\Delta x \to 0$. Using Theorem 5.1, we have

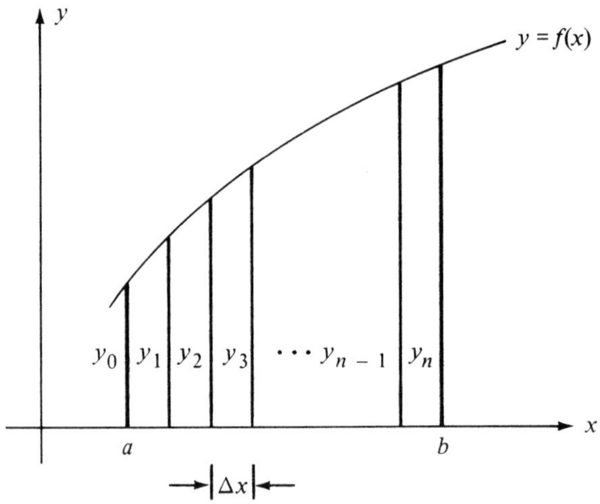

Fig. 11.1

$$\bar{y} = \frac{1}{b-a}\int_a^b y\,\mathrm{d}x.$$

When dealing with periodic functions a useful parameter is the **root-mean-square (RMS) value**. For a function $f(t)$ of period T the RMS *value* is defined as

$$\sqrt{\frac{1}{T}\int_0^T \{f(t)\}^2\,\mathrm{d}t}\,,$$

i.e. the *square root* of the *mean value* of the *function squared* taken over an interval equal to *one period.*

Example 11.1

(i) Find the mean value of the function $x^2 + 1$ in the interval $1 \leqslant x \leqslant 3$.

(ii) Find the RMS value of a 50 Hz voltage signal of amplitude 250 V.

(i) Let $y = x^2 + 1$.

$$\bar{y} = \frac{1}{3-1}\int_1^3 (x^2+1)\,\mathrm{d}x$$

$$= \tfrac{1}{2}\left[\frac{x^3}{3} + x\right]_1^3$$

$$= \tfrac{16}{3}.$$

(ii) The frequency of 50 Hz is $50 \times 2\pi = 100\pi$ rad/s, i.e. $V = 250 \sin 100\pi t$ and the period T is 1/50 s. Therefore,

$$\text{mean square value} = \frac{1}{0.02}\int_0^{0.02} 250^2 \sin^2 100\pi t \, dt$$

$$= 50 \int_0^{0.02} \frac{250^2}{2}(1 - \cos 200\pi t)\, dt$$

$$= 25 \times 250^2 \left[t - \frac{1}{200\pi}\sin 200\pi t\right]_0^{0.02}$$

$$= 25 \times 250^2 \times 0.02$$

$$= 0.5 \times 250^2.$$

$$\text{RMS value} = \sqrt{0.5 \times 250^2}$$

$$= \frac{250}{\sqrt{2}}\text{ V}.$$

Problems

1. Find the mean value of the following functions.

   (i) $y = \dfrac{1}{1+x^2}, \quad -1 \leqslant x \leqslant 1.$

   (ii) $y = \sin 2t, \ 0 \leqslant t \leqslant \dfrac{\pi}{2}.$

2. Find the RMS value of the following periodic functions.
   (i) $y = \cos t \cos 2t$ (period, 2π).
   (ii) $y = 1 + \sin 3t$.
   (iii) $y = t - 2n, \quad -1 + 2n \leqslant t < 1 + 2n, \quad n = 0, \pm 1, \pm 2, \ldots$.
   (Hint: sketch the curve for $-3 \leqslant t < 3$, i.e. $n = -1, 0, 1$.)

3. Show that the RMS value of $A \sin \omega t$ where A and ω are constants is $A/\sqrt{2}$.

Answers

1. (i) $\dfrac{\pi}{4}$.

   (ii) $\dfrac{2}{\pi}$.

2. (i) $\frac{1}{2}$.

(ii) $\sqrt{\dfrac{3}{2}}$.

(iii) $\dfrac{1}{\sqrt{3}}$.

11.3 LENGTH OF A CURVE AND AREA OF A SURFACE

Suppose that it is required to calculate the length of the curve $y = f(x)$ between two given points $P(x_1, y_1)$ and $Q(x_2, y_2)$. Divide the curve up into a number of sections, the arc length of a typical section being denoted by ds. The projections of ds parallel to the coordinate axes are dx and dy (Fig. 11.2).

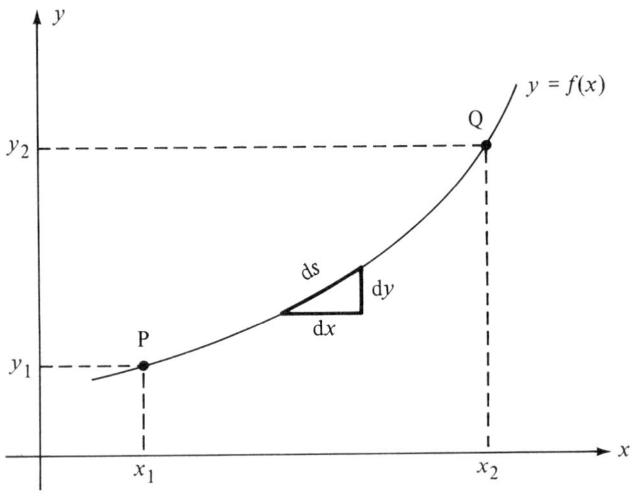

Fig. 11.2

If ds is small we have approximately

$$(ds)^2 = (dx)^2 + (dy)^2,$$

i.e.

$$ds = \sqrt{(dx)^2 + (dy)^2}$$

$$= \sqrt{1 + \left(\frac{dy}{dx}\right)^2}\, dx.$$

Applying Theorem 5.1, the required arc length is $\int ds$ evaluated between the points P and Q, i.e.

$$\text{arc length} = \int_{x_1}^{x_2} \sqrt{1 + \left(\frac{dy}{dx}\right)^2}\, dx. \tag{11.1}$$

Alternatively, we can write

$$ds = \sqrt{1 + \left(\frac{dx}{dy}\right)^2}\, dy$$

or

$$ds = \sqrt{\left(\frac{dx}{dt}\right)^2 + \left(\frac{dy}{dt}\right)^2}\, dt,$$

so that

$$\text{arc length} = \int_{y_1}^{y_2} \sqrt{1 + \left(\frac{dx}{dy}\right)^2}\, dy \qquad (11.2)$$

or

$$\text{arc length} = \int_{t_1}^{t_2} \sqrt{\left(\frac{dx}{dt}\right)^2 + \left(\frac{dy}{dt}\right)^2}\, dt, \qquad (11.3)$$

where t_1 and t_2 are the parametric values associated with the points P and Q.

The choice between the three versions given in equations (11.1)–(11.3) will depend on the way in which the equation of the curve is specified (see Example 11.2).

If the area under the curve $y = f(x)$ between $x = x_1$ and $x = x_2$ is rotated about the x axis to form a solid, the *curved surface area* of the solid may be obtained by considering the area swept out by the element ds of arc and applying Theorem 5.1, i.e.

$$\text{curved surface area} = \int 2\pi y\, ds$$

evaluated between P and Q. Any of the three versions for ds may be used depending on the specification of the curve. In all cases it is necessary to draw a sketch and to use the correct radius of rotation (do not try to remember formulae).

Example 11.2

(i) Find the length of the curve $y = \frac{1}{3}(x^2 + 2)^{3/2}$ between the points where $x = 0$ and $x = 3$.

(ii) Use the parametric form of the equation of a circle, of radius a (i.e. $x = a \cos t$, $y = a \sin t$) to confirm the result for the circumference of a circle.

(iii) Find the length of the curve $9x^2 = 4y^3$ between the points $(0, 0)$ and $(2\sqrt{3}, 3)$.

(i) $y = \frac{1}{3}(x^2 + 2)^{3/2}$,

$$\frac{dy}{dx} = x(x^2 + 2)^{1/2}.$$

Therefore, use

$$ds = \sqrt{1 + \left(\frac{dy}{dx}\right)^2}\ dx$$

$$= \sqrt{1 + x^2(x^2 + 2)}\ dx$$

$$= \sqrt{(x^2 + 1)^2}\ dx$$

$$= (x^2 + 1)\ dx$$

and

$$\text{arc length} = \int_0^3 (x^2 + 1)\ dx$$

$$= 12 \text{ units.}$$

(ii) $$x = a \cos t,$$

$$y = a \sin t.$$

$$\frac{dx}{dt} = -a \sin t,$$

$$\frac{dy}{dt} = a \cos t.$$

Therefore, use

$$ds = \sqrt{\left(\frac{dx}{dt}\right)^2 + \left(\frac{dy}{dt}\right)^2}\ dt$$

$$= \sqrt{a^2 \sin^2 t + a^2 \cos^2 t}\ dt$$

$$= \sqrt{a^2}\ dt$$

$$= a\ dt$$

and

$$\text{arc length} = \int_0^{2\pi} a\ dt$$

$$= 2\pi a.$$

(iii) $9x^2 = 4y^3$,

$$x = \tfrac{2}{3} y^{3/2},$$

$$\frac{dx}{dy} = y^{1/2}.$$

$$ds = \sqrt{1 + \left(\frac{dx}{dy}\right)^2}\; dy$$

$$= \sqrt{1+y}\; dy.$$

Therefore,

$$\text{arc length} = \int_0^3 \sqrt{1+y}\; dy$$

$$= [\tfrac{2}{3}(1+y)^{3/2}]_0^3$$

$$= \tfrac{14}{3} \text{ units.}$$

(Note that the integral would have been much more difficult if the formulation

$$\int_0^{2\sqrt{3}} \sqrt{1 + \left(\frac{dy}{dx}\right)^2}\; dx$$

had been used.)

Example 11.3

(i) Find the curved surface area of a sphere by rotating the circle $x^2 + y^2 = a^2$ about the x axis.

(ii) Find the curved surface area generated when the arc of the curve $y = x^2$ between $(0, 0)$ and $(2, 4)$ is rotated about the y axis.

(i) We write the circle equation in parametric form

$$x = a \cos t,$$

$$y = a \sin t,$$

and use

$$ds = \sqrt{\left(\frac{dx}{dt}\right)^2 + \left(\frac{dy}{dt}\right)^2}\; dt$$

$$= \sqrt{a^2 \sin^2 t + a^2 \cos^2 t}\; dt = a\, dt.$$

The radius of rotation is y (Fig. 11.3(a)), Therefore,

$$\text{curved surface area} = \int_{t=0}^{t=\pi} 2\pi y\; ds$$

$$= \int_0^{\pi} 2\pi\, a \sin t\, a\, dt$$

$$= 2\pi a^2 [-\cos t]_0^{\pi}$$

$$= 4\pi a^2.$$

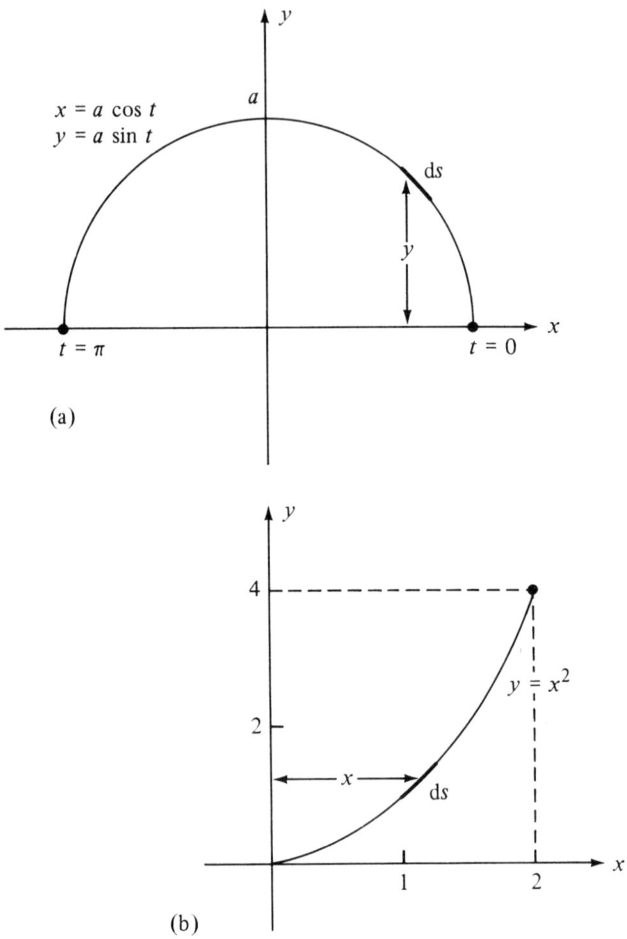

Fig. 11.3

(ii) $y = x^2$,

$$\frac{dy}{dx} = 2x.$$

The radius of rotation is x (Fig. 11.3(b)). Use

$$ds = \sqrt{1 + \left(\frac{dy}{dx}\right)^2}\ dx$$

$$= \sqrt{1 + 4x^2}\ dx.$$

Therefore,

$$\text{curved surface area} = \int_{x=0}^{x=2} 2\pi x \, ds$$

$$= \int_0^2 2\pi x \sqrt{1 + 4x^2} \, dx.$$

Substituting $u = 1 + 4x^2$ we obtain

$$\text{curved surface area} = \int_1^{17} \frac{\pi}{4} u^{1/2} \, du$$

$$= \frac{\pi}{6}(17^{3/2} - 1) \text{ units}^2.$$

(Confirm that the same result is obtained more readily from

$$\int_0^4 2\pi x \sqrt{1 + \left(\frac{dx}{dy}\right)^2} \, dy = \int_0^4 \pi\sqrt{4y + 1} \, dy.)$$

Problems

1. Calculate the following arc lengths.
   (i) The length of the curve

   $$y = \frac{x^3}{3} + \frac{1}{4x}$$

   between $x = 1$ and $x = 3$.

   (ii) The length of the curve

   $$x = \frac{y^4}{4} + \frac{1}{8y^2}$$

   between $y = 1$ and $y = 2$.

2. Show that the arc length of the curve $9x^2 = (2y + 3)^3$ between the points $(\sqrt{3}, 0)$ and $(9, 3)$ is $\frac{1}{3}(10\sqrt{10} - 8)$.

3. (i) Find the arc length of the curve $x = \frac{1}{2}t^2$, $y = \frac{1}{3}(2t + 1)^{3/2}$ between $t = 0$ and $t = 4$.

   (ii) A curve is defined in parametric form by

   $$x = a \cos t + at \sin t,$$

   $$y = a \sin t - at \cos t, \qquad 0 \leqslant t \leqslant \frac{\pi}{2}.$$

   Show that the arc length is $\frac{1}{8}a\pi^2$.

4. A solid is formed by rotating the area under $y = x^3$ about the x axis between $x = 0$ and $x = 1$. Find the volume and curved surface area of the solid.

5. (i) The arc length in problem 1 (ii) is rotated about the x axis. Calculate the curved surface area generated.
   (ii) The arc length in problem 1 (i) is rotated about the line $y = -1$. Calculate the curved surface area generated.

6. (i) The curve given by

$$x = t + 1,$$
$$y = \tfrac{1}{2}t^2 + t, \quad 0 \leqslant t \leqslant 4,$$

is rotated about the y axis. Show that the curved surface area generated is

$$\frac{2\pi}{3}(26\sqrt{26} - 2\sqrt{2}).$$

   (ii) The closed curve $x = a\cos^3\theta$, $y = a\sin^3\theta$ is rotated about the x axis. Show that the curved surface area generated is

$$\frac{12\pi a^2}{5}.$$

Answers

1. (i) $\frac{53}{6}$.
   (ii) $\frac{123}{32}$.

3. (i) 12.

4. $\dfrac{\pi}{7}$, $\dfrac{\pi}{27}(10\sqrt{10} - 1)$.

5. (i) $\dfrac{253\pi}{20}$.
   (ii) $\dfrac{1823\pi}{18}$.

11.4 CENTROID AND CENTRE OF GRAVITY

The *centroid* of a plane figure is a point whose position depends on the shape of the figure. It is very much like the concept of the *centre of gravity* of a thin plate but we cannot call it 'centre of gravity' because a plane figure has no mass. In fact, if a thin plate is of *uniform density* and has the *same shape* as the plane figure, then the centre of gravity of the thin plate and the centroid of the plane figure are in corresponding positions.

To define the position of the centroid of a plane figure with respect to coordinate axes Oxy, consider the area A shown in Fig. 11.4.

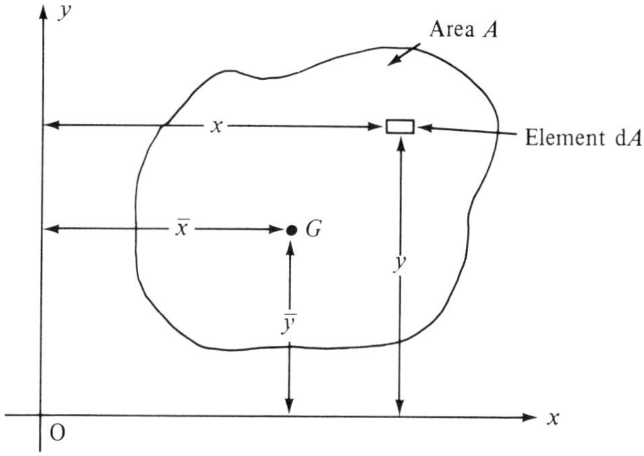

Fig. 11.4

The area is divided into a large number of *small* elements with a typical element of area dA, situated at (x, y), shown in the figure. G$(\bar{x}, \bar{y})$ indicates the position of the *centroid*.

We define the **first moment of area** of the element dA about the y axis as x dA and the first moment of area about Oy for the *entire area A* is obtained by summing the first moments of all the elements. We denote the first moment of A about Oy by $\int x \, \mathrm{d}A$. This definition of first moment of area about the y axis is still valid when the small element dA is modified to be a *thin* strip parallel to the y axis.

The coordinate $\bar{x}$ is defined such that the first moment of A about Oy is also $A\bar{x}$, i.e.

$$A\bar{x} = \int x \, \mathrm{d}A$$

where

$$A = \int \mathrm{d}A,$$

the area of the plane figure.

By defining first moments about the x axis in a similar way, we obtain the result

$$A\bar{y} = \int y \, \mathrm{d}A,$$

In this case the element dA may be a *thin* strip parallel to the x axis. It follows that the coordinates of the centroid of the plane figure are given by

$$\bar{x} = \frac{\int x \, dA}{\int dA}, \qquad (11.4)$$

$$\bar{y} = \frac{\int y \, dA}{\int dA}. \qquad (11.5)$$

When applying equation (11.4) *directly* (i.e. by integration), it is important to construct the element dA such that all parts of it are the same distance from y axis. Similarly, equation (11.5) can only be applied when all parts of dA are the same distance from the x axis. In other situations, special care must be exercised.

The application of equations (11.4) and (11.5) is demonstrated in Example 11.4 and Example 11.5.

When calculating the position of a centroid the following short cuts are useful.

(i) The centroid of a *regular figure* is at its geometrical centre. (For example the centroid of a rectangle is at the point where the diagonals cross.)
(ii) The centroid always lies on an *axis of symmetry*.
(iii) For *composite figures* the total first moments can be obtained by adding the first moments of each part of the figure (see Example 11.6).

Example 11.4
Derive the formulae

$$\bar{x} = \frac{\int_a^b xy \, dx}{\int_a^b y \, dx} \qquad (11.6)$$

and

$$\bar{y} = \frac{\frac{1}{2}\int_a^b y^2 \, dx}{\int_a^b y \, dx} \qquad (11.7)$$

for the centroid $(\bar{x}, \bar{y})$ of the area under the curve $y = f(x), f(x) \geqslant 0$. between $x = a$ and $x = b$.

Apply the formulae to calculate the position of the centroid of the area under the curve $y = x^2 + 2$ between $x = 0$ and $x = 2$.

The area under $y = f(x)$ is divided into strips drawn parallel to the y axis, each strip of width dx. A typical strip is shown in Fig. 11.5 and this strip is taken to be the element of area dA. Since dx is small, the area of the element is given approximately by

$$dA = y \, dx.$$

The centroid of the *element* is at $(x, y/2)$. Taking first moments of area about the y axis and summing over all the strips,

$$A\bar{x} = \int_{x=a}^{x=b} x \, dA = \int_a^b xy \, dx,$$

where A is the area under the curve given by

$$A = \int_a^b y \, dx.$$

Therefore,

$$\bar{x} = \frac{\int_a^b xy \, dx}{\int_a^b y \, dx}.$$

Taking first moments of area about the x axis and summing over all the strips,

$$A\bar{y} = \int_{x=a}^{x=b} \frac{y}{2} \, dA = \tfrac{1}{2} \int_a^b y^2 \, dx.$$

Therefore,

$$\bar{y} = \frac{\tfrac{1}{2} \int_a^b y^2 \, dx}{\int_a^b y \, dx}.$$

It should be noted that equation (11.6) follows immediately from equation (11.4) by writing $dA = y \, dx$. This is because *all parts of the thin element* dA *are the same distance from the* y *axis*. This is not the case for the x axis and we must be careful when taking moments about the x axis; equation (11.7) *cannot* be obtained from equation (11.5) by replacing dA by $y \, dx$.

Applying equations (11.6) and (11.7) to the area under $y = x^2 + 2, 0 \leqslant x \leqslant 2$,

$$A = \int_0^2 (x^2 + 2) \, dx$$

$$= \left[\frac{x^3}{3} + 2x\right]_0^2$$

$$= \tfrac{20}{3}.$$

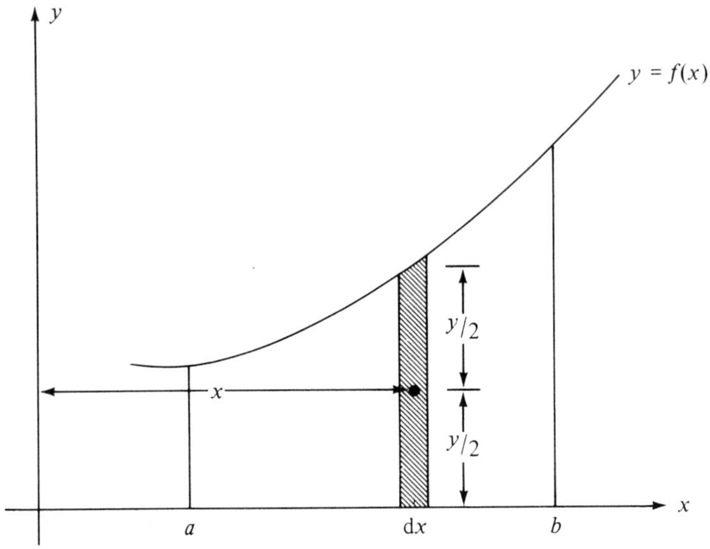

Fig. 11.5

$$\bar{x} = \tfrac{3}{20}\int_0^2 x(x^2+2)\,\mathrm{d}x$$

$$= \tfrac{3}{20}\left[\frac{x^4}{4}+x^2\right]_0^2$$

$$= \tfrac{6}{5}.$$

$$\bar{y} = \tfrac{3}{20}\,\tfrac{1}{2}\int_0^2 (x^2+2)^2\,\mathrm{d}x$$

$$= \tfrac{3}{40}\int_0^2 (x^4+4x^2+4)\,\mathrm{d}x$$

$$= \tfrac{3}{40}\left[\frac{x^5}{5}+\frac{4x^3}{3}+4x\right]_0^2$$

$$= \tfrac{47}{25}.$$

Therefore the centroid of the area is at (1.2, 1.88).

It is important to understand that the equations (11.6) and (11.7) only apply to the particular case of the area under a curve. It is advisable to work from first principles in all other situations.

Example 11.5

Show that the centroid of a triangle is at a distance of one-third of its height from the base.

The triangular area is divided into thin strips parallel to the base (Fig. 11.6).

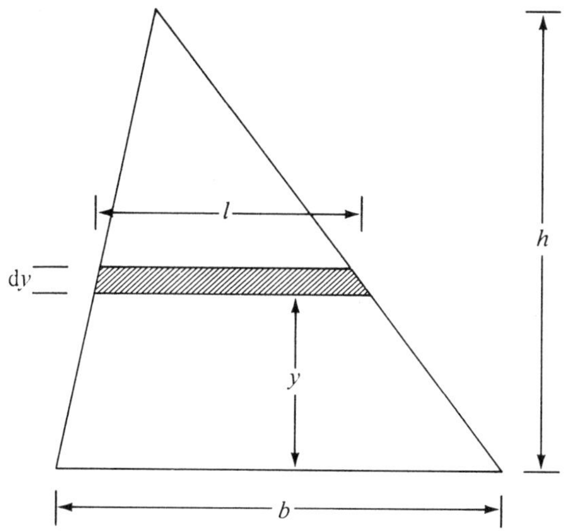

Fig. 11.6

Each strip has width dy and the typical strip is at distance y from the base. The strip width l is given by similar triangles as

$$l = \frac{b}{h}(h - y).$$

The area of the strip is $\mathrm{d}A = l\,\mathrm{d}y$ and, taking first moments of area of the triangle about its base

$$A\bar{y} = \int_0^h y(l\,\mathrm{d}y),$$

where $A = \frac{1}{2}bh$ is the area of the triangle and $\bar{y}$ is the distance of the centroid G from the base.

Therefore,

$$\bar{y} = \frac{2}{bh}\int_0^h y\frac{b}{h}(h - y)\,\mathrm{d}y$$

$$= \frac{2}{h^2}\int_0^h (hy - y^2)\,\mathrm{d}y$$

$$= \frac{2}{h^2}\left[\frac{hy^2}{2} - \frac{y^3}{3}\right]_0^h$$

$$= \frac{h}{3},$$

which proves the required result.

Example 11.6

Fig. 11.7(a) shows the cross-section of a beam. Calculate the position of the centroid of the section.

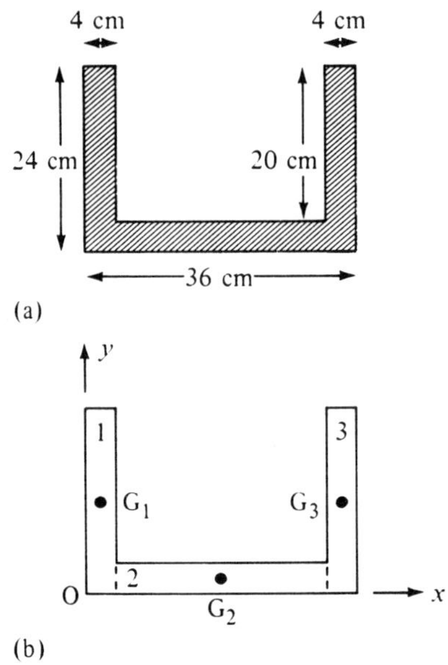

Fig. 11.7

Divide the section into three rectangles and choose axes Oxy as shown in Fig. 11.7(b). Rectangle 1 is 4×24 and its centroid G_1 is at (2, 12). Rectangle 2 is 28×4 and its centroid G_2 is at (18, 2). Rectangle 3 is 4×24 and its centroid G_3 is at (34, 12).

Let $G(\bar{x}, \bar{y})$ be the position of the centroid of the section.

Clearly, $x = 18$ is an axis of symmetry of the figure and therefore $\bar{x} = 18$.

Taking first moments of area about the x axis, we obtain

$$A\bar{y} = (4 \times 24) \times 12 + (28 \times 4) \times 2 + (4 \times 24) \times 12$$
$$= 2528 \text{ cm}^3,$$

where A is the section area given by

$$A = (4 \times 24) + (28 \times 4) + (4 \times 24)$$
$$= 304 \text{ cm}^2.$$

Therefore,

$$\bar{y} = \frac{2528}{304}$$
$$\approx 8.316 \text{ cm},$$

i.e. the centroid of the section is at (18, 8.316) relative to the axes Oxy.

Example 11.7

Determine the position of the centre of gravity of a solid right circular cone of uniform density ρ kg/m^3.

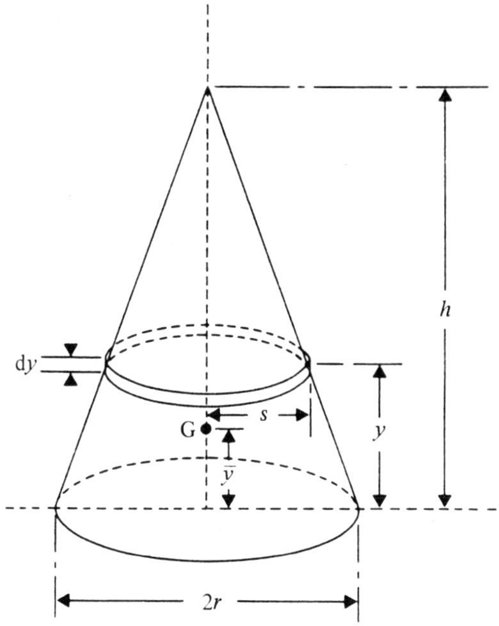

Fig. 11.8

Let h m and r m be the height and base radius of the cone. The cone is divided into elements parallel to its base and of thickness dy. Each element is thus thin enough to be considered as a thin circular cylinder or *disc*. A typical disc is shown in Fig. 11.8 and its radius s is determined by similar triangles as

$$s = \frac{r}{h}(h - y).$$

The volume of the disc is $\pi s^2 \, dy$ and its mass is $\rho \pi s^2 \, dy$. The centre of gravity of the disc is at distance y from the base of the cone and, taking first moments of mass about a base diameter,

$$M\bar{y} = \int_0^h y(\rho \pi s^2 \, dy),$$

where $M = \frac{1}{3}\rho\pi r^2 h$ is the mass of the cone and $\bar{y}$ is the distance of the centre of gravity of the cone from its base. (G lies on the axis of symmetry of the cone.) Therefore,

$$\bar{y} = \frac{3}{\rho\pi r^2 h}\int_0^h \rho\pi y \frac{r^2}{h^2}(h-y)^2\,dy$$

$$= \frac{3}{h^3}\int_0^h (h^2y - 2hy^2 + y^3)\,dy$$

$$= \frac{3}{h^3}\left[\frac{h^2y^2}{2} - \frac{2hy^3}{3} + \frac{y^4}{4}\right]_0^h$$

$$= \frac{h}{4},$$

i.e. the centre of gravity of a cone is situated at a distance of one-quarter of the height from the base.

Problems

1. Find the centroid of the area enclosed by the part of the curve $y = 4 - x^2$ which lies above the x axis.

2. Find the centroid of the area bounded by the curves $y^2 = 4x$ and $x^2 = 4y$.

3. Find the position of the centroid of the section shown in Fig. 11.9 with respect to the axes given.

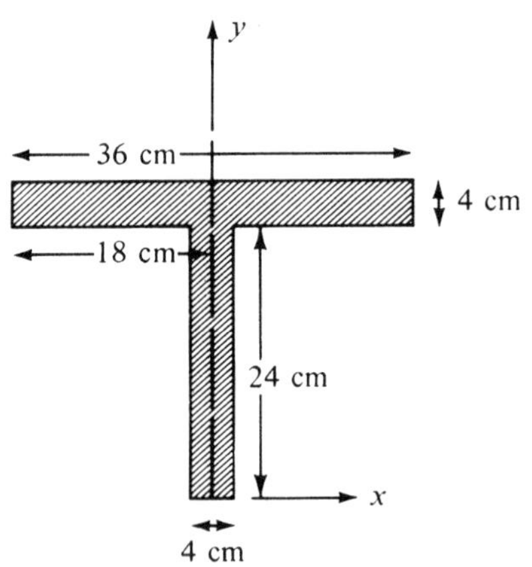

Fig. 11.9

4. Show that the centroid of a trapezium with parallel sides of length a and b, of distance h apart, is at a point of distance

$$\frac{h(2a+b)}{3(a+b)}$$

from the side of length b.

5. Prove that the centroid of the region bounded by the curve $y^2 = 4x$ and the line $x = b$, $b \geqslant 0$, is the point $(3b/5, 0)$.

6. Show that the distance of the centroid of a semicircular disc of radius r is $4r/3\pi$ from the diameter.

7. Determine the distance of the centre of gravity of a uniform solid hemisphere from its base.

8. A non-uniform rod of length 5 m has a *linear* density at a point distance x from one end given by

$$\rho = \left(2 + \frac{x}{20}\right)^2 \text{ kg/m.}$$

Find the following.
(i) The mass of the rod.
(ii) The distance of the centre of gravity from the end of the rod.

Answers
1. (0, 1.6).
2. (1.8, 1.8).
3. (0, 20.4).
7. Three-eights of the radius.
8. (i) 22.60 kg.
(ii) 2.60 m from the lighter end.

11.5 SECOND MOMENT OF AREA AND MOMENT OF INTERTIA

In section 11.4 we defined the first moment of area of a plane figure about an axis. Referring again to Fig. 11.4, we define the **second moment of area** of the element dA about the y axis at x^2 dA. Similarly the second moment of area of dA about the x axis is y^2 dA.

The second moment of area for the *entire* area A are obtained by summing the second moments of all the elements. Denoting the second moment of area about the x axis by I_x and the second moment of area about the y axis by I_y, we have

$$I_x = \int y^2 \, \mathrm{d}A, \tag{11.8}$$

$$I_y = \int x^2 \, \mathrm{d}A, \tag{11.9}$$

where as before the integral sign has been used to represent the summation and the limiting process of Theorem 5.1. When applying equation (11.8) or equation (11.9), it is essential that the element is constructed so that *all parts of it are the same distance* from the axis under consideration.

It is sometimes convenient to write the second moment of area I of a plane figure about an axis in the form $I = Ak^2$, where A is the area of the figure. k has the dimensions of length and is called the **radius of gyration** of the area with respect to the axis.

An important application of second moments of area is in structural problems associated with the bending of beams. For example, the following formula is used in beam analysis:

$$\frac{\sigma}{y} = \frac{M}{I} = \frac{E}{R},$$

where σ is the stress, y is the distance from the axis through the centroid of the section, M is the bending moment, I is the second moment of area of the beam section about the axis through the centroid, E is Young's modulus and R is the radius of curvature.

Closely associated with the second moment of area is the concept of the **moment of inertia** which is a property of solids. The moment of inertia of a solid about a given axis is defined as $\int r^2 \, dm$ where dm is an *element of mass* and r is the distance of the element from the axis. Again the element must be such that all parts of it are the same distance from the axis, and the integral sign represents summation over the entire solid. If the moment of inertia of the solid is written in the form $I = Mk^2$ where M is the mass of the solid, then k is called the *radius of gyration* of the solid with respect to the axis. For the particular case when the solid is a thin flat plate of *uniform surface density* ρ (mass per unit area), then the moment of inertia of the plate about the axis is given by

$$\begin{aligned} I &= \int r^2 \, dm \\ &= \int r^2 \rho \, dA \\ &= \rho \int r^2 \, dA \end{aligned}$$

since ρ is constant. Thus,

$$I = \rho \times \text{second moment of area of the shape about the axis.}$$

Also since $I = Mk^2 = \rho Ak^2$, it follows that the second moment of area of the plate area about the axis is Ak^2, i.e. the radius of gyration is the *same value* whether defined in terms of the moment of inertia or in terms of the second moment of area.

Moments of inertia are applied when studying rotating dynamical systems for which Newtons' second law takes the form

$$T = I \frac{d^2 \theta}{dt^2},$$

where T is the applied torque, I is the moment of intertia about the axis of rotation and $d^2\theta/dt^2$ is the angular acceleration. (Note that the units for second moments of area are *length*^{4} while the units for moments of intertia are *mass* $\times$ *length*^{2}.)

The following theorems simplify the calculation of second moments of area and with certain restrictions may be applied to moments of inertia.

Theorem 11.1 (Parallel axes theorem)
If I_x is the second moment of area of a plane figure about axis XX′ and I_g is the second moment of area about a parallel axis GG′ through the *centroid*, then

$$I_x = I_g + Ad^2$$

where A is the area of the figure and d is the distance between the parallel axes GG′ and XX′.

Proof

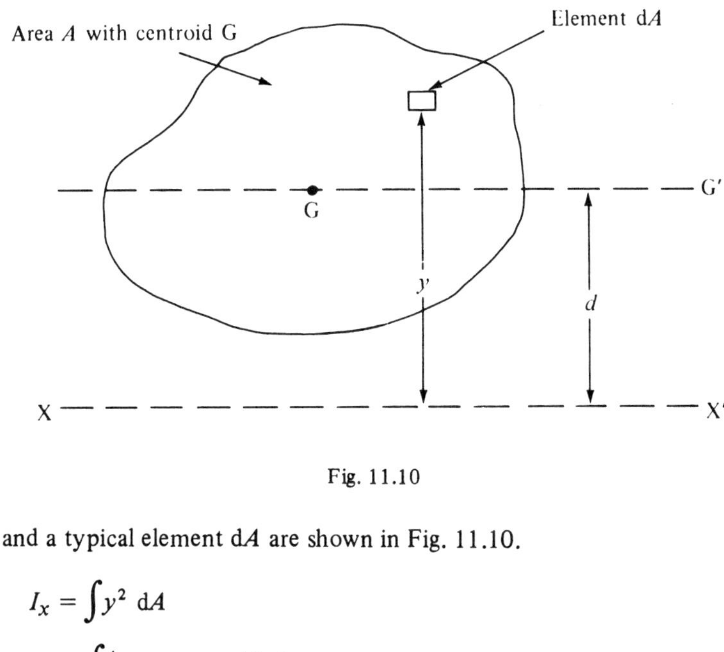

Fig. 11.10

The area and a typical element dA are shown in Fig. 11.10.

$$\begin{aligned} I_x &= \int y^2 \, dA \\ &= \int \{d + (y - d)\}^2 \, dA \\ &= \int d^2 \, dA + 2\int d(y-d) \, dA + \int (y-d)^2 \, dA. \end{aligned}$$

The first term is Ad^2 since d is a constant. The second term is a multiple of the *first* moment of area about an axis through the centroid which is zero by definition of the centroid. The third term is I_g, the second moment of area about GG′. Therefore,

$$I_x = I_g + Ad^2.$$

A simple extension of the proof shows that the theorem can be applied to the moment of inertia of a *solid* in the form

$$I_x = I_g + Md^2,$$

where I_x and I_g now refer to moments of inertia about XX′ and GG′ and M is the mass of the solid.

Theorem 11.2 (Perpendicular axes theorem)
If I_x and I_y are the second moments of area of a plane figure about perpendicular axes Ox and Oy in the plane of the figure, and I_z is the second moment of area about an axis through O, *perpendicular to the plane of the figure*, then

$$I_z = I_x + I_y.$$

Proof

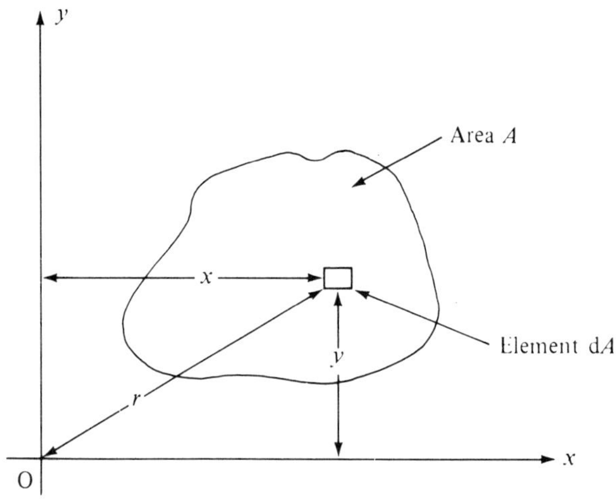

Fig. 11.11

Fig. 11.11 shows the plane area A and a typical element of dA. The axis Oz is through O perpendicular to the figure.

$$\begin{aligned} I_z &= \int r^2 \, \mathrm{d}A \\ &= \int (x^2 + y^2) \, \mathrm{d}A \\ &= \int x^2 \, \mathrm{d}A + \int y^2 \, \mathrm{d}A \\ &= I_y + I_x. \end{aligned}$$

Theorem 11.2 may also be applied to the calculation of the moment of inertia of a thin flat plate about an axis perpendicular to the plane of the plate.

Examples (11.8) and (11.9) are used to derive standard results for the second moments of areas of rectangles and circles. Examples (11.10)–(11.12) apply these results to more general areas. The methods for calculating moments of inertia for a solid of circular section are demonstrated in Example 11.13.

Example 11.8

(i) Calculate the second moment of area of a $b \times d$ rectangle about an axis containing a side of length b.

(ii) Calculate the second moment of area of the rectangle about a parallel axis through the centroid of the rectangle.

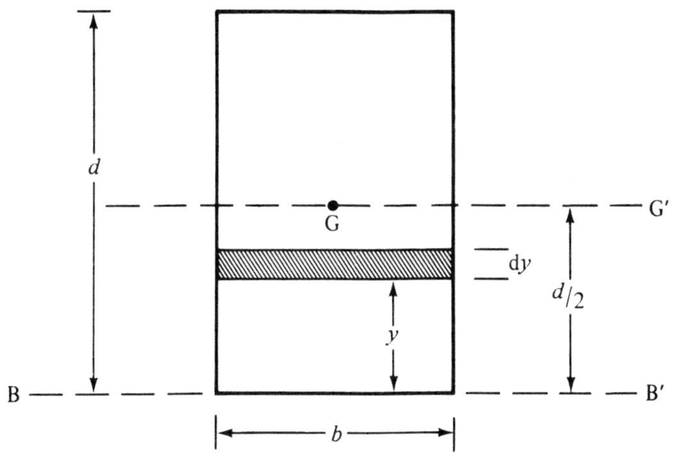

Fig. 11.12

Fig. 11.12 shows the rectangle and the required axes BB′ and GG′. The element of area is a thin strip of depth dy (chosen so that all parts are equidistant from the axes). Let I_b and I_g denote the required second moments of area.

(i) Applying equation (11.8),

$$
\begin{aligned}
I_b &= \int_0^d y^2 \, \mathrm{d}A \\
&= \int_0^d y^2 b \, \mathrm{d}y \\
&= b\left[\frac{y^3}{3}\right]_0^d \\
&= \frac{bd^3}{3} .
\end{aligned}
$$

(ii) Applying Theorem 11.1,

$$I_b = I_g + bd\left(\frac{d}{2}\right)^2$$

Therefore,

$$I_g = \frac{bd^3}{3} - \frac{bd^3}{4}$$

$$= \frac{bd^3}{12}.$$

Example 11.9

(i) Determine the radius of gyration for a circular area of radius a with respect to an axis through its centre perpendicular to the plane of the circle.

(ii) Deduce that the radius of gyration of the circular area with respect to an axis containing a diameter is $a/2$.

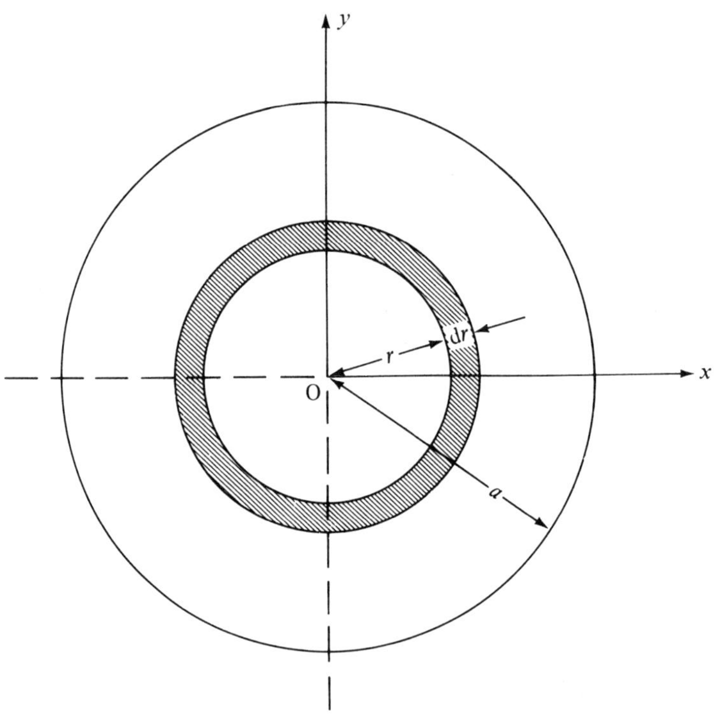

Fig. 11.13

(i) The element of area is a thin ring of width dr (chosen so that all parts of it are equidistant from the perpendicular axis) (Fig. 11.13). The area of the element is approximately $dA = 2\pi r\,dr$ and the second moment I_z of area about the perpendicular axis is obtained from equation (11.8):

$$I_z = \int r^2 \, dA$$

$$= \int_0^a r^2 \, 2\pi r \, dr$$

$$= 2\pi \left[\frac{r^4}{4} \right]_0^a$$

$$= \frac{\pi a^4}{2}$$

$$= A \left(\frac{a}{\sqrt{2}} \right)^2$$

where $A = \pi a^2$, the area of the circle. Therefore the radius of gyration of the circular area with respect to the perpendicular axis is $a/\sqrt{2}$.

(ii) Denoting the second moment of area of the circular area about Ox and Oy by I_x and I_y respectively, we have that $I_x = I_y$, both axes are diameters, and

$$I_z = \frac{\pi a^4}{2} = I_x + I_y$$

from Theorem 11.2. Therefore,

$$I_x = I_y = \frac{\pi a^4}{4} = \pi a^2 \left(\frac{a}{2} \right)^2.$$

i.e. the radius of gyration with respect to a diameter is $a/2$.

Example 11.10

Determine the second moment of area of the T section shown in Fig. 11.14(a) about the axes Gx and Gy where G is the centroid of the section.

From problem 3 in section 11.4 the centroid G of the figure is 20.4 cm above the base of the figure.

Divide the section into two rectangles as shown in Fig. 11.14(b). We apply Theorem 11.1 to each of the rectangles to obtain their second moments about Gx.

For rectangle 1 the second moment of area about an axis through its centroid G_1 parallel to Gx is

$$\frac{36 \times 4^3}{12}$$

(Example 11.8 (ii)). Therefore the second moment of rectangle 1 about axis Gx is

$$\frac{36 \times 4^3}{12} + (36 \times 4) \times (5.6)^2 = 4707.84 \text{ cm}^4.$$

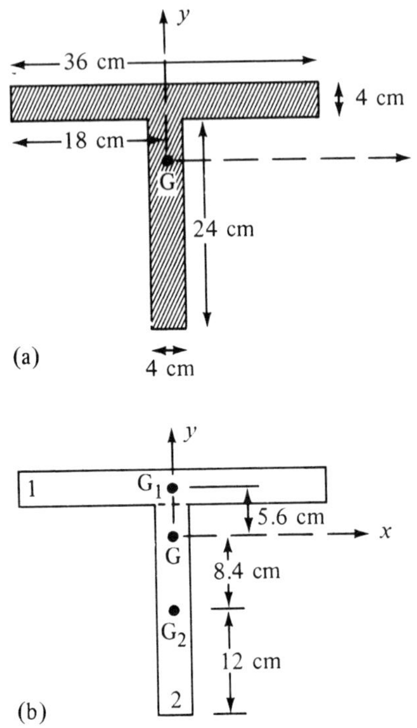

Fig. 11.4

Similarly the second moment of area of rectangle 2 about Gx is

$$\frac{4 \times 24^3}{12} + (4 \times 24) \times (8.4)^2 = 11\,381.76 \text{ cm}^4.$$

Therefore the second moment of the T section about Gx is

$$4707.84 + 11\,381.76 = 16\,089.6 \text{ cm}^4.$$

The axis Gy passes through both G_1 and G_2, the centroids of the two rectangles. Therefore we may apply Example 11.8 (ii) directly for each rectangle, adding the results. The second moment of the T section about Gy is

$$\frac{4 \times 36^3}{12} + \frac{24 \times 4^3}{12} = 15\,680 \text{ cm}^4.$$

Example 11.11
Find the second moment of area of a semicircle of radius a about its diameter.

A complete circle is made up of two semicircles each of which has the same second moment of area about a diameter. Therefore the second moment of area of a semicircle about a diameter is half of the value for the complete circle, i.e.

$$\frac{1}{2}\frac{\pi a^4}{4} = \frac{\pi a^4}{8}.$$

(Note that this can be written

$$\frac{\pi a^2}{2}\left(\frac{a}{2}\right)^2$$

and so the radius of gyration about the diameter is $a/2$ which is the *same* as for the complete circle.)

Example 11.12

For the area under the curve $y = f(x)$ between $x = a$ and $x = b$, show that the second moments of area about the x axis and y axis are given respectively by

$$I_x = \tfrac{1}{3}\int_a^b y^3 \, dx$$

and

$$I_y = \int_a^b x^2 y \, dx.$$

The area under the curve is divided into strips drawn parallel to the y axis as shown in Fig. 11.5. Each strip is approximately a rectangle of area $y\,dx$ whose second moment about the base is $dx\, y^3/3$. Since the base of the strip lies in the x axis, the second moment of the area about the x axis is obtained by summation over all the strips and is given by

$$I_x = \tfrac{1}{3}\int_a^b y^3 \, dx.$$

When taking second moments about the y axis the *thin* strip qualifies as an element (all parts are same distance from the axis). Therefore, equation (11.9) can be applied directly, i.e.

$$I_y = \int x^2 \, dA$$

$$= \int_a^b x^2 y \, dx.$$

Example 11.13

Calculate the moment of inertia in the form Mk^2 for a uniform circular cylinder

(i) about its axis and

(ii) about an axis through its centroid perpendicular to the axis of the cylinder.

Let ρ (mass per unit volume) be the density of the cylinder.

Let l be the length of the cylinder and a the radius of its cross-section. Choose axes Gxy as shown in Fig. 11.15 where G is the centroid of the cylinder. The cylinder is divided into thin discs of thickness dx; a typical disc is shown in the figure.

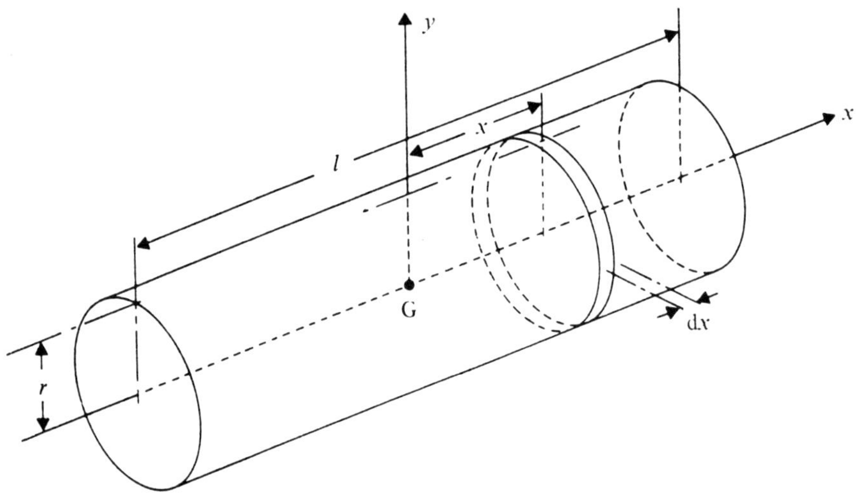

Fig. 11.15

(i) The thin disc can be considered as a flat plate of uniform density and therefore its radius of gyration with respect to the axis Gx is $a/\sqrt{2}$ (Example 11.9 (i)). The mass of the disc is $\rho\pi a^2\,\mathrm{d}x$ and its moment of inertia about Gx is

$$\rho\pi a^2\,\mathrm{d}x\left(\frac{a}{\sqrt{2}}\right)^2 = \frac{\rho\pi a^4}{2}\,\mathrm{d}x.$$

Therefore the moment of inertia of the cylinder about Gx is

$$\int_{-l/2}^{l/2}\frac{\rho\pi a^4}{2}\,\mathrm{d}x = \frac{\rho\pi a^4}{2}\,[x]_{-l/2}^{l/2}$$

$$= \frac{\rho\pi a^4 l}{2}$$

$$= M\left(\frac{a}{\sqrt{2}}\right)^2$$

where $M = \rho\pi a^2 l$ is the mass of the cylinder.

(ii) The radius of gyration of the disc with respect to its diameter (parallel to the axis Gy) is $a/2$ (Example 11.9 (ii)). Therefore the moment of inertia of the disc about this diameter is

$$\rho\pi a^2\ \mathrm{d}x\left(\frac{a}{2}\right)^2 = \frac{\rho\pi a^4}{4}\ \mathrm{d}x.$$

Applying Theorem 11.1, the moment of inertia of the disc about axis Gy is

$$\frac{\rho\pi a^4}{4}\ \mathrm{d}x + \rho\pi a^2\ \mathrm{d}x\, x^2 = \rho\pi a^2\left(\frac{a^2}{4} + x^2\right)\ \mathrm{d}x.$$

Therefore the moment of inertia of the cylinder about Gy is

$$\int_{-l/2}^{l/2} \rho\pi a^2\left(\frac{a^2}{4} + x^2\right)\ \mathrm{d}x = \rho\pi a^2\left[\frac{a^2 x}{4} + \frac{x^3}{3}\right]_{-l/2}^{l/2}$$

$$= 2\rho\pi a^2\left(\frac{a^2 l}{8} + \frac{l^3}{24}\right)$$

$$= \rho\pi a^2 l\left(\frac{a^2}{4} + \frac{l^2}{12}\right)$$

$$= M\left(\sqrt{\frac{a^2}{4} + \frac{l^2}{12}}\right)^2.$$

Problems

1. Find the second moment of area of a rectangular section of dimensions $b \times d$, about the following.
   (i) An axis through its centroid perpendicular to the plane of the section.
   (ii) An axis through one corner of the section parallel to the axis in (i).

2. Calculate the second moment of area of a square of side a about a diagonal.

3. Show that the second moment of area of a triangle, of base b and height h, about an axis containing its base is $bh^3/12$. Determine the second moment of area of the triangle about a parallel axis
   (i) through the centroid of the triangle and
   (ii) through the vertex of the triangle.

4. A circular door of a boiler is hinged so that it turns about a tangent. If its area is 0.503 m^2, find the radius of gyration about the hinge axis.

5. (i) Find the second moment of area of the section shown in Fig. 11.16 about the axis XX$'$.
   (ii) Find the second moment of area of the beam section shown in Fig. 11.17 about the axis XX$'$ and about a parallel axis through the centroid of the section.

6. Calculate the second moments of area about the coordinate axes of the plane area bounded by the curve $y = x^2 + 3$, the x axis and the lines $x = 1$ and $x = 3$.

7. The depth of the centre of pressure of a vertical area immersed in liquid is given by

$$\bar{\bar{z}} = \frac{\text{second moment of area}}{\text{first moment of area}},$$

where both moments are taken about the axis lying in the surface of the liquid and in the plane of the immersed area. Find $\bar{\bar{z}}$ in the following cases.
   (i) A vertical dam 40 m × 20 m, the top edge of dam (40 m) lying in the surface.
   (ii) A vertical dam in the shape of a trapezium with parallel sides a and b at a distance D apart.
   (iii) The immersed area shown in Fig. 11.18.

8. Determine the moment of inertia of a uniform sphere of radius a about a diameter.

9. Determine the moment of inertia of a uniform cone of height h and base radius r for the following.
   (i) About the axis of the cone.
   (ii) About an axis through the vertex of the cone parallel to the base.

Answers

1. (i) $\dfrac{bd}{12}(b^2 + d^2)$.

   (ii) $\dfrac{bd}{3}(b^2 + d^2)$.

2. $\dfrac{a^4}{12}$.

3. (i) $\dfrac{bh^3}{36}$.

   (ii) $\dfrac{bh^3}{4}$.

4. 0.447 m.
5. (i) 56 cm^4.
   (ii) 396 cm^4, 75.67 cm^4.
6. $I_x = 345.3$ units^4, $I_y = 74.4$ units^4.
7. (i) 13.33 m.

   (ii) $\dfrac{d(a + 3b)}{2(a + 2b)}$.

   (iii) 14.3 m.

8. $\frac{2}{5}Ma^2$ where M is the mass of the sphere.
9. (i) $\frac{3}{10}Mr^2$.
   (ii) $\frac{3}{20}M(r^2 + 4h^2)$ where M is the mass of the cone.

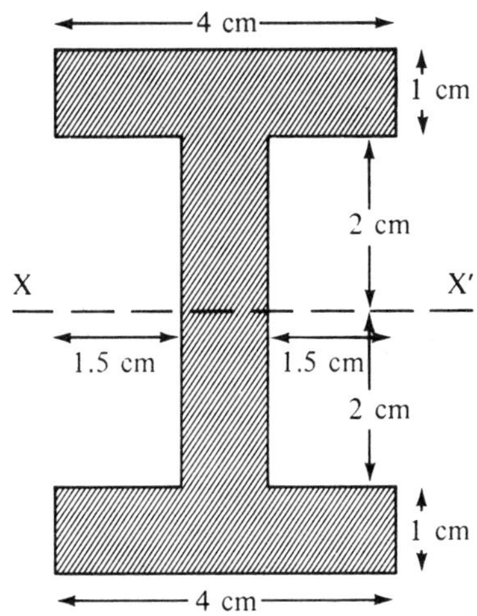

Fig. 11.16

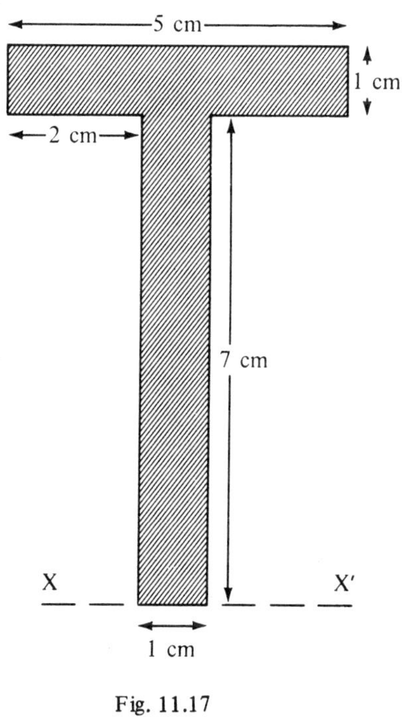

Fig. 11.17

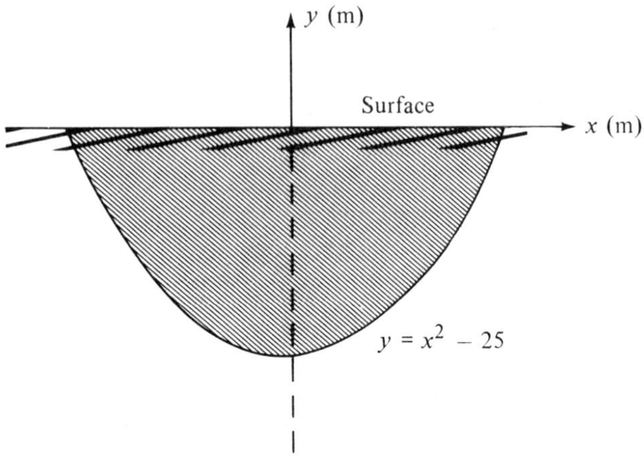

Fig. 11.18

Further Problems for Chapter 11

1. Determine the mean value of $f(x) = \cos^2 x$ in the interval $0 \leqslant x \leqslant \pi/2$.

2. Calculate the RMS value of the function $1 + \cos t$.

3. Write down a definite integral which is equal to the length of the curve $y^2 = 4x$ between the points (0, 0) and (1, 2).

4. The arc of the curve $y^2 = 4x$ between $x = 1$ and $x = 3$, is rotated about the y axis to generate a surface. Show that the area of one side of this surface is

$$2\pi \int_1^3 \sqrt{x^2 + x}\, dx.$$

5. The area enclosed between the curves $y = \sqrt{x}$ and $y = x$ between (0, 0) and (1, 1) is given by

$$\int_0^1 (\sqrt{x} - x)\, dx = \tfrac{1}{6} \text{ units}^2.$$

Write down in terms of definite integrals the position of the centroid of the area.

6. Express in terms of definite integrals the second moments of area of the area in the previous example about the coordinate axes.

Answers

1. $\frac{1}{2}$.
2. $\sqrt{1.5}$.

3. $\int_0^1 \sqrt{1+\frac{1}{x}}\,\mathrm{d}x$ or $\int_0^2 \sqrt{1+\frac{y^2}{4}}\,\mathrm{d}y$.

5. $\bar{x} = 6\int_0^1 (x\sqrt{x} - x^2)\,\mathrm{d}x$, $\bar{y} = 3\int_0^1 (x - x^2)\,\mathrm{d}x$.

6. $I_y = \int_0^1 (x^2\sqrt{x} - x^3)\,\mathrm{d}x$, $I_x = \int_0^1 (y^3 - y^4)\,\mathrm{d}y$.

12

Approximate Integration

12.1 INTRODUCTION

There are many simple functions which possess no indefinite integral in terms of elementary functions (elementary functions are the types already encountered in this textbook). For example the functions $\sin(x^2)$ and $\sqrt{1+x^3}$ can not be integrated without defining new functions, i.e. no combination of elementary functions can be differentiated to obtain either $\sin(x^2)$ or $\sqrt{1+x^3}$. However, it may be necessary to calculate the *definite* integral of such a function as the result of one of the applications studied in Chapter 5 and Chapter 11.

We recall that

$$\int_a^b f(x)\,dx$$

is equal to the area under the curve $y = f(x)$ between $x = a$ and $x = b$ (assuming that $f(x) \geqslant 0, a \leqslant x \leqslant b$). Therefore, we can approximate to the value of

$$\int_a^b f(x)\,dx$$

by *measuring* this area in some way. The area may be obtained approximately by either graphical or numerical techniques. A simple graphical technique is to draw an accurate graph of $y = f(x)$ and to estimate the area under the curve by counting squares on the graph paper. Two numerical methods are considered in this chapter, the *trapezoidal rule* and the more accurate *Simpson's rule*. In Chapter 13,

$$\int_a^b f(x)\,dx$$

is estimated by an alternative technique which requires the expansion of $f(x)$ as an infinite series.

12.2 THE TRAPEZOIDAL RULE AND SIMPSON'S RULE

The area under the curve $y = f(x)$, $a \leqslant x \leqslant b$, is divided into n strips of equal width w as shown in Fig. 12.1.

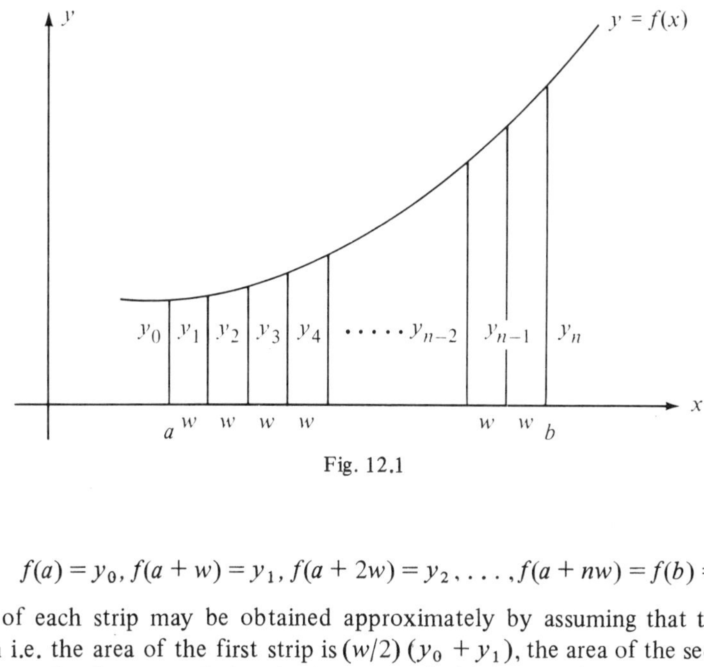

Fig. 12.1

Let

$$f(a) = y_0, f(a + w) = y_1, f(a + 2w) = y_2, \ldots, f(a + nw) = f(b) = y_n.$$

The area of each strip may be obtained approximately by assuming that the strip is a trapezium i.e. the area of the first strip is $(w/2)(y_0 + y_1)$, the area of the second strip is $(w/2)(y_1 + y_2)$, the area of the third strip is $(w/2)(y_2 + y_3)$, etc. The area under $y = f(x)$, $a \leqslant x \leqslant b$, is given approximately by the sum of the areas of the n trapezia. Since the area is given *exactly* by

$$\int_a^b f(x)\,dx,$$

we have

$$\int_a^b f(x)\,dx \approx \frac{w}{2}\{y_0 + 2(y_1 + y_2 + y_3 + \ldots y_{n-1}) + y_n\}. \tag{12.1}$$

Equation (12.1) is the **trapezoidal rule** for the approximate evaluation of a definite integral. The accuracy of the approximation may be improved by increasing the number of strips. It can be shown that the *error* made by using equation (12.1) to evaluate

$$\int_a^b f(x)\,dx$$

will not exceed the maximum value of $|(w^2/12)(b-a)f''(x)|$ in $a \leqslant x \leqslant b$ (see Phillips and Taylor (1973)).

It is clear that the trapezoidal rule will give the *exact* value for

$$\int_a^b f(x)\,dx$$

when $f(x)$ is a linear function of x and, in this case, only *one* strip is required. More accurate integration formulae may be derived by ensuring that they give exact results for functions which are polynomials of higher degree.

For example, we attempt to determine a formula which is exact when $f(x)$ is a third-degree polynomial and when *two* strips are used.

Consider the area under the curve

$$y = px^3 + qx^2 + rx + s, \quad -w \leqslant x \leqslant w.$$

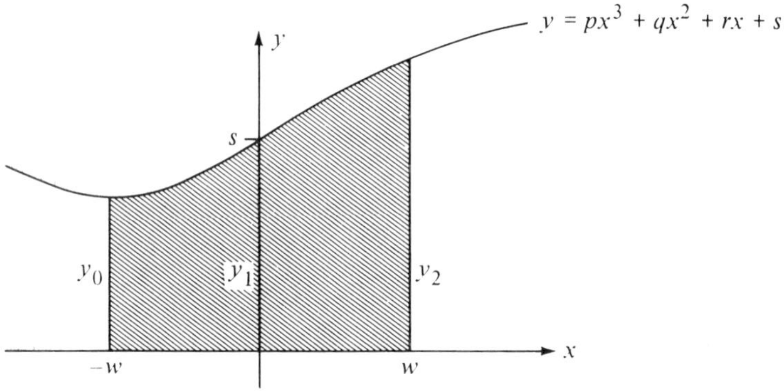

Fig. 12.2

The area is shown in Fig. 12.2; y_0, y_1 and y_2 are the ordinates for the x values $-w$, 0 and w, i.e.

$$y_0 = -pw^3 + qw^2 - rw + s, \tag{12.2}$$

$$y_1 = s \tag{12.3}$$

and

$$y_2 = pw^3 + qw^2 + rw + s. \tag{12.4}$$

The exact value of the area is given by

$$\int_{-w}^{w} (px^3 + qx^2 + rx + s)\,dx = \left[\frac{px^4}{4} + \frac{qx^3}{3} + \frac{rx^2}{2} + sx\right]_{-w}^{w}$$

$$= \frac{2qw^3}{3} + 2sw$$

$$= w\left(\frac{2qw^2}{3} + 2s\right).$$

To obtain an integration formula of a type similar to equation (12.1), we attempt to find three numbers A, B and C such that the area under the curve is $w(Ay_0 + By_1 + Cy_2)$, i.e. we require

$$w(Ay_0 + By_1 + Cy_2) = w\left(\frac{2qw^2}{3} + 2s\right). \tag{12.5}$$

Substituting for y_0, y_1 and y_2 from equations (12.2)–(12.4) and equating coefficients of w in equation (12.5), we obtain

$$-pA + pC = 0,$$

$$qA + qC = \frac{2q}{3},$$

$$As + Bs + Cs = 2s,$$

i.e. $A = C = \frac{1}{3}$ and $B = \frac{4}{3}$. We have, for $f(x) = px^3 + qx^2 + rx + s$,

$$\int_{-w}^{w} f(x)\,dx = \frac{w}{3}(y_0 + 4y_1 + y_2). \tag{12.6}$$

This result is not dependent on the position of the y axis and therefore applies for *any* area under a third-degree polynomial curve. The formula will also give exactly the areas under linear and quadratic curves. For any other curve $y = f(x)$, an approximate value of the area is obtained by applying equation (12.6).

Returning to the evaluation of

$$\int_a^b f(x)\,dx,$$

we can apply equation (12.6) provided that the area under the curve in $a \leqslant x \leqslant b$ is divided into an *even* number of strips. Taking the strips in pairs and referring again to Fig. 12.1, we have

$$\text{area of first pair} \approx \frac{w}{3}(y_0 + 4y_1 + y_2),$$

$$\text{area of second pair} \approx \frac{w}{3}(y_2 + 4y_3 + y_4),$$

$$\text{area of third pair} \approx \frac{w}{3}(y_4 + 4y_5 + y_6), \qquad \text{etc.}$$

Therefore, summing over all the strips, we have

$$\int_a^b f(x)\,\mathrm{d}x \approx \frac{w}{3}\{y_0 + 4(y_1 + y_2 + \ldots + y_{n-1})$$
$$+ 2(y_2 + y_4 + \ldots + y_{n-2}) + y_n\}. \tag{12.7}$$

When $f(x)$ is a third-degree (or lower-degree) polynomial, equation (12.7) will be exact; only two strips are required (i.e. use equation (12.6)).

Equation (12.7) is known as **Simpson's rule**. The error made by using Simpson's rule to evaluate

$$\int_a^b f(x)\,\mathrm{d}x$$

will not exceed the maximum value of $|(w^4/180)\,(b-a)\,f''''(x)|$ in $a \leqslant x \leqslant b$ (see Phillips and Taylor (1973)).

It should be clear that the restriction $f(x) \geqslant 0$ may be relaxed when using equations (12.1), (12.6) and (12.7) to evaluate the definite integral

$$\int_a^b f(x)\,\mathrm{d}x.$$

Example 12.1

Find an approximate value for

$$\int_0^2 \sqrt{1+x^3}\,\mathrm{d}x$$

using the following.

(i) The trapezoidal rule with $n = 4$.
(ii) Simpson's rule with $n = 4$.
(iii) Simpson's rule with $n = 10$.

(i) $n = 4$, $w = 0.5$.

x	$\sqrt{1+x^3}$		Multiplier
0	$1 = y_0$		1
0.5		$1.0606 = y_1$	2
1.0		$1.4142 = y_2$	2
1.5		$2.0916 = y_3$	2
2.0	$3 = y_4$		1

$y_1 + y_2 + y_3 = 4.5664.$

Therefore,

$$\int_0^2 \sqrt{1+x^3}\,dx = \frac{0.5}{2}(1 + 2 \times 4.5664 + 3)$$
$$\approx 3.283.$$

(ii) $n = 4$, $w = 0.5$.

x		$\sqrt{1+x^3}$		Multiplier
0	$1 = y_0$			1
0.5		$1.0606 = y_1$		4
1.0			$1.4142 = y_2$	2
1.5		$2.0916 = y_3$		4
2.0	$3 = y_4$			1

$$y_1 + y_3 = 3.1522.$$

Therefore,

$$\int_0^2 \sqrt{1+x^3}\,dx = \frac{0.5}{3}(1 + 4 \times 3.1522 + 2 \times 1.4142 + 3)$$
$$\approx 3.240.$$

(iii) $n = 10$, $w = 0.2$.

x		$\sqrt{1+x^3}$		Multiplier
0	$1 = y_0$			1
0.2		$1.0039 = y_1$		4
0.4			$1.0315 = y_2$	2
0.6		$1.1027 = y_3$		4
0.8			$1.2296 = y_4$	2
1.0		$1.4142 = y_5$		4
1.2			$1.6517 = y_6$	2
1.4		$1.9349 = y_7$		4
1.6			$2.2574 = y_8$	2
1.8		$2.6138 = y_9$		4
2.0	$3 = y_{10}$			1

$y_1 + y_3 + y_5 + y_7 + y_9 = 8.0695,$

$y_2 + y_4 + y_6 + y_8 = 6.1702.$

Therefore,

$$\int_0^2 \sqrt{1+x^3}\,\mathrm{d}x = \frac{0.2}{3}(1 + 4 \times 8.0695 + 2 \times 6.1702 + 3)$$

$$\approx 3.241.$$

Example 12.2
Use Simpson's rule to confirm the result for the volume of a hemisphere of radius a, i.e. $V = \frac{2}{3}\pi a^3$.

The hemisphere is generated by rotating the quarter-circle $y = \sqrt{a^2 - x^2}$, $0 \leqslant x \leqslant a$, about the x axis. Applying the method of discs described in Chapter 5, we obtain for the volume of the hemisphere

$$V = \int_0^a \pi(a^2 - x^2)\,\mathrm{d}x.$$

Since the integrand, $\pi(a^2 - x^2)$, is a quadratic function of x, Simpson's rule will give the *exact* value of the volume. Only two strips are required and equation (12.6) may be applied.

Denoting the cross-section area $\pi(a^2 - x^2)$ by $A(x)$ and defining A_0 and A_2 as the 'end-section' areas and A_1 as the 'mid-section' area, i.e.

$$A_0 = A(0) = \pi a^2,$$

$$A_1 = A\left(\frac{a}{2}\right) = \frac{3\pi a^2}{4},$$

$$A_2 = A(a) = 0,$$

equation (12.6) gives

$$V = \frac{1}{3}\,\frac{a}{2}\,(A_0 + 4A_1 + A_2)$$

$$= \frac{a}{6}(\pi a^2 + 3\pi a^2)$$

$$= \frac{2\pi a^3}{3}.$$

The formula

$$\text{volume} = \tfrac{1}{6} \times \text{length} \times (A_0 + 4A_1 + A_2) \qquad (12.8)$$

may be used to determine the approximate volume of an irregular solid. It will, of course, give the exact volume when the cross-section area is a third-degree (or lower-degree) polynomial function of distance from one end.

Example 12.3

A tunnel 100 m long has a square cross-section. The height of the tunnel at distance x from one end is shown in the table. Use the trapezoidal rule to estimate the volume of earth removed during the construction of the tunnel.

x (m)	0	10	20	30	40	50	60	70	80	90	100
Height (m)	4	3.83	3.71	3.80	3.92	3.97	4.04	4.21	4.31	4.23	4.17

If the cross-section area at distance x is $A(x)$, then the volume of a 'slice' of thickness dx is $A(x)\,dx$ and the total volume is

$$\int_0^{100} A(x)\,dx.$$

We tabulate $A(x)$ versus x and apply equation (12.1).

x	0	10	20	30	40	50	60	70
$A(x)$	16	14.6689	13.7641	14.4400	15.3664	15.7609	16.3216	17.7241
x	80	90	100					
$A(x)$	18.5761	17.8929	17.3889					

$$\sum_{n=1}^{9} A(10n) = 144.5150.$$

Therefore,

$$\begin{aligned}\text{volume} &\approx \frac{10}{2}(16 + 2 \times 144.5150 + 17.3889)\\ &\approx 1612\ \text{m}^3.\end{aligned}$$

A quick approximation to the amount of earth removed may be obtained by applying equation (12.8). $A_0 = 16, A_1 = 15.7609, A_2 = 17.3889$ and

$$\begin{aligned}\text{volume} &\approx \frac{100}{6}(16 + 4 \times 15.7609 + 17.3889)\\ &\approx 1607\ \text{m}^3.\end{aligned}$$

Problems

1. Obtain an approximation to

$$\int_0^1 e^{-x^2}\,dx$$

using the following methods.
(i) Trapezoidal rule with $n = 5$.
(ii) Trapezoidal rule with $n = 10$.
(iii) Simpson's rule with $n = 2$.
(iv) Simpson's rule with $n = 10$.
(Give your answers correct to three decimal places.)

2. Use Simpson's rule to confirm the formula for the volume of the cap of a sphere, of radius R:

$$V = \tfrac{1}{3}\pi h^2(3R - h),$$

where h is the height of the cap.

3. The areas, A m^2, of the surface of water in a reservoir, when the depth of water above a reference level is h m, are given in the table. Estimate the volume of water above the reference level when $h = 30$ m using the following methods.
(i) The trapezoidal rule.
(ii) Simpson's rule.

h	0	5	10	15	20	25	30
$\dfrac{A}{100}$	718	740	800	850	893	900	900

4. The motion of a certain body in a resisting medium satisfies the equation

$$v\frac{\mathrm{d}v}{\mathrm{d}x} = -(10 + 0.01v^2)$$

and the condition $v = 10$ when $x = 0$, where v is the velocity of the body at a distance x from a fixed point. Show that

$$x = \int_v^{10} \frac{u\,\mathrm{d}u}{10 + 0.01u^2}.$$

(Hint: obtain $\mathrm{d}x/\mathrm{d}v$.) Calculate correct to three decimal places the value of x when $v = 0$ using the trapezoidal rule, choosing the strip width to ensure the required accuracy. Check your answer using integration by substitution.

5. A point P is at a distance h from the plane which contains a figure of area A. A solid is generated by drawing straight lines from P to all points on the perimeter of the figure (i.e. the cross-sections of the solid taken parallel to the figure are of similar shape). Show that the volume of the solid is $Ah/3$.

Answers

1. (i) 0.744.
   (ii) 0.746.
   (iii) 0.747.
   (iv) 0.747.
3. (i) 2 496 000 m^3.
   (ii) 2 494 000 m^3.
4. 4.766.

Further Problems for Chapter 12

1. Use equation (12.1) with five strips to estimate the value of

$$\int_0^1 x^2 \, dx$$

and compare your answers with the exact value.

2. If equation (12.7) with four strips was applied to the integral in the previous question, what result would you expect?

3. Use the result of problem 5 in section 12.2 to obtain the formula for the volume of a cone. Confirm your formula by applying Simpson's rule.

4. What size would you expect the error to be when applying Simpson's rule with four strips to the calculation of

$$\int_0^2 x^4 \, dx?$$

5. Check the result of the previous example by calculation of the exact and approximate values of

$$\int_0^2 x^4 \, dx.$$

Answers

1. The estimated value is 0.34; the exact value is $\frac{1}{3}$.
2. The exact value.
4. $\frac{1}{60}$.

13

Infinite Series

13.1 INTRODUCTION

In this chapter we attempt to express a given function $f(x)$ as an infinite series in ascending powers of x, i.e.

$$f(x) = a_0 + a_1x + a_2x^2 + a_3x^3 + \ldots,$$

where $a_0, a_1, a_2, a_3, \ldots$ are constants.

Such a series is known as a *power series*.

In summation notation we have

$$f(x) = \sum_{i=0}^{\infty} a_i x^i.$$

The motivation for expressing $f(x)$ in this way is that when $f(x)$ is small we may be able to find a *polynomial approximation* to the function by *terminating* the power series after an appropriate number of terms.

For example, we shall show that the function $\sin x$ may be expressed as the power series

$$\sin x = x - \frac{x^3}{3!} + \frac{x^5}{5!} - \frac{x^7}{7!} + \ldots.$$

We have already seen that $\sin x = x$ is a good approximation when x is small (i.e. the above series is terminated after one term). A better approximation is obtained by including the next term of the series, i.e.

$$\sin x = x - \frac{x^3}{6}.$$

This third-degree polynomial approximation to the function $\sin x$ is only about 1.2% in error for a value of x as large as $\pi/3$.

Using polynomials as approximations is valuable because they are the simplest class of functions. Polynomials are readily evaluated and are easy to differentiate and integrate as many times as is necessary.

The series given above for $\sin x$ is a valid representation for the function for *all* values of x. This is frequently *not* the case for a power series representation of a given function. For example, consider the well-known *geometric series*

$$\frac{1}{1-x} = 1 + x + x^2 + x^3 + \ldots . \tag{13.1}$$

For $x = 0.1$ the left-hand side of equation (13.1) has the value $\frac{10}{9}$, and the right hand side is $1.1111\ldots = \frac{10}{9}$. Therefore, equation (13.1) is valid for $x = 0.1$. For $x = 2$, however, the left-hand side is -1 whereas the right-hand side, $1 + 2 + 4 + 8 + \ldots$, is undefined and equation (13.1) is clearly not valid. In fact the infinite series

$$1 + x + x^2 + x^3 + \ldots$$

is a valid representation of the function

$$\frac{1}{1-x}$$

only for the interval $-1 < x < 1$. This is easily proved if we recall the formula for the sum of a *finite* geometric series:

$$S_n = \sum_{i=0}^{n-1} x_i = \frac{1-x^n}{1-x}.$$

As $n \to \infty$,

$$S_n \to \sum_{i=0}^{\infty} x^i = \frac{1}{1-x}$$

provided that $-1 < x < 1$; otherwise, $\lim_{n\to\infty} S_n$ does not exist.

Equation (13.1) is an *incomplete statement*; it is only useful when the condition $-1 < x < 1$ is added. Any approximation based on a termination of the series will be invalid if x does not lie in the interval $-1 < x < 1$. We say that the series is *convergent* for $-1 < x < 1$. Convergence is studied briefly in section 13.2 and a simple test is stated.

In section 13.3, *Maclaurin's theorem* is used to determine the power series representation for a given function (provided that the series exists). Applications to the evaluation of definite integrals are considered.

13.2 CONVERGENCE OF AN INFINITE SERIES

Corresponding to the power-series representation

$$\sum_{i=0}^{\infty} a_i x^i$$

of the function $f(x)$ we define the function of n

$$S(n) = \sum_{i=0}^{n-1} a_i x^i;$$

$S(n)$ is the sum of the first n terms of the power series. The power series is a valid representation of the function $f(x)$ for a particular value of x provided that $\lim_{n \to \infty} [S(n)]$ exists and is equal to $f(x)$. If this is the case, then the power series is said to **converge** to $f(x)$; if the limit does not exist, the series is said to be **divergent.**

For a given series

$$\sum_{i=0}^{\infty} a_i x^i,$$

there will be a number $R > 0$ such that the series converges for $-R < x < R$ and diverges for $|x| > R$. The interval $-R < x < R$ is called the **interval of convergence** and R is called the **radius of convergence.**

When expressing a function $f(x)$ as an infinite series in ascending powers of x, it is *essential* that the interval of convergence is also stated so that invalid use of the series can be avoided.

A detailed study of convergence is beyond the scope of this book; for more information, see Dunning-Davies (1982). Fortunately there is a simple method for determination of the interval of convergence of a given series without first having to obtain the function $S(n)$. The method depends on Theorem 13.1 which is stated without proof.

Theorem 13.1 (The ratio test for convergence)

The infinite series

$$\sum_{i=0}^{\infty} u_i \qquad u_i \neq 0 \text{ for all } i,$$

is convergent provided that

$$\lim_{n \to \infty} \left| \frac{u_{n+1}}{u_n} \right| < 1$$

and is divergent if

$$\lim_{n \to \infty} \left| \frac{u_{n+1}}{u_n} \right| > 1.$$

(If the limit is exactly 1 then the series *may or may not* be convergent.)

The application of this theorem to infinite series in ascending powers of x is demonstrated in Example 13.1. Problem 1 in this section deals with the power-series expansions of certain common functions.

Example 13.1

(i) Determine the interval of convergence for the infinite series

$$1+\frac{x}{2}+\frac{2x^2}{2^2}+\frac{3x^3}{2^3}+\frac{4x^4}{2^4}+\ldots.$$

(ii) Confirm that the series

$$\sin x = x-\frac{x^3}{3!}+\frac{x^5}{5!}-\frac{x^7}{7!}+\ldots.$$

is convergent for all values of x.

(i) Let

$$u_1=\frac{x}{2},$$

$$u_2=\frac{2x^2}{2^2}.$$

$$u_3=\frac{3x^3}{2^3},$$

etc.

(Note that it does not matter which term of the series is called u_1.) By inspection,

$$|u_n|=\left|\frac{nx^n}{2^n}\right|$$

and

$$|u_{n+1}|=\left|\frac{(n+1)x^{n+1}}{2^{n+1}}\right|$$

(simply replace n by $n+1$). Therefore,

$$\left|\frac{u_{n+1}}{u_n}\right|=\left|\frac{(n+1)x^{n+1}}{2^{n+1}}\quad\frac{2^n}{nx^n}\right|$$

$$=\frac{n+1}{2n}\,|x|.$$

$$\lim_{n\to\infty}\left|\frac{u_{n+1}}{u_n}\right| = \tfrac{1}{2}|x|$$

and by Theorem 13.1 the given series is convergent when $\frac{1}{2}|x| < 1$, and divergent when $\frac{1}{2}|x| > 1$, i.e. the series is convergent for $-2 < x < 2$ and divergent for $|x| > 2$. (The theorem does not apply when $x = 2$ or -2 but it is clear from inspection of the series that it is divergent for these values of x.)

(ii) Let

$$u_1 = x,$$

$$u_2 = -\frac{x^3}{3!},$$

$$u_3 = \frac{x^5}{5!},$$

etc.

By inspection,

$$|u_n| = \left|\frac{x^{2n-1}}{(2n-1)!}\right|$$

and

$$|u_{n+1}| = \left|\frac{x^{2n+1}}{(2n+1)!}\right|.$$

Therefore,

$$\left|\frac{u_{n+1}}{u_n}\right| = \left|\frac{x^{2n+1}}{(2n+1)!}\quad\frac{(2n-1)!}{x^{2n-1}}\right|$$

$$= \frac{x^2}{(2n+1)2n}.$$

$$\lim_{n\to\infty}\left|\frac{u_{n+1}}{u_n}\right| = 0$$

and, applying Theorem 13.1, the series for $\sin x$ will be convergent when $0 < 1$, which is true for *all* values of x.

Problems

1. For each of the following power-series expansions obtain the largest value of R such that the series is convergent for $-R < x < R$ (i.e. determine the radius of convergence).

(i) $\cos x = 1 - \frac{x^2}{2!} + \frac{x^4}{4!} - \frac{x^6}{6!} + \ldots$.

(ii) $\tan^{-1} x = x - \frac{x^3}{3} + \frac{x^5}{5} - \frac{x^7}{7} + \ldots$.

(iii) $\ln(1 + x) = x - \frac{x^2}{2} + \frac{x^3}{3} - \frac{x^4}{4} + \ldots$.

(iv) $e^x = 1 + x + \frac{x^2}{2!} + \frac{x^3}{3!} + \ldots$.

(v) $\sinh x = x + \frac{x^3}{3!} + \frac{x^5}{5!} + \frac{x^7}{7!} + \ldots$.

(vi) $(1 + x)^n = 1 + nx + \frac{n(n-1)}{2!} x^2 + \frac{n(n-1)(n-2)}{3!} x^3 + \ldots$,

where n is not a positive integer.

2. Explain why the series for $(1 + x)^n$ given in problem 1 (vi) is valid for *all* x when n is a positive integer.

Answers

1. (i) ∞.
   (ii) 1.
   (iii) 1.
   (iv) ∞.
   (v) ∞.
   (vi) 1.
2. The series is finite with $n + 1$ terms.

13.3 MACLAURIN'S SERIES FOR THE FUNCTION $f(x)$

For a given function $f(x)$ we can obtain its infinite-series expansion in ascending powers of x (provided that the series exists) by application of Theorem 13.2.

Theorem 13.2 (Maclaurin's theorem)

$$f(x) = f(0) + f'(0)x + \frac{f''(0)}{2!}x^2 + \frac{f'''(0)}{3!}x^3 + \ldots \qquad (13.2)$$

provided that

(i) $f(0), f'(0), f''(0), \ldots$ all exist, and
(ii) the value of x is such that series converges to $f(x)$.

Equation (13.2) is known as the **Maclaurin's series** for the function $f(x)$.

In the summation notation, equation (13.2) can be written

$$f(x) = \sum_{r=0}^{\infty} \frac{f^{(r)}(0)}{r!} x^r,$$

where $f^{(r)}(0)$ denotes the value of $d^r f/dx^r$ when $x = 0$.

The formal proof of Theorem 13.2 is not given here but, if we assume that a series exists of the form

$$f(x) = a_0 + a_1 x + a_2 x^2 + a_3 x^3 + \dots ,$$

then differentiation r times of this assumed form followed by replacement of x by zero confirms that

$$a_r = \frac{f^{(r)}(0)}{r!}.$$

Equation (13.2) gives a representation of $f(x)$ which is *useful near* $x = 0$ since, if x is sufficiently small (within the interval of convergence), the series may be terminated to provide a polynomial approximation for $f(x)$. If we require a series which is useful near a *non-zero* value of x, $x = a$, then we require a more general form of equation (13.2):

$$f(x) = f(a) + f'(a)(x - a) + \frac{f''(a)}{2!}(x - a)^2 + \frac{f'''(a)}{3!}(x - a)^3 + \dots . \qquad (13.3)$$

Equation (13.3) is known as the **Taylor's series** for the function $f(x)$.

For example, the function

$$f(x) = \frac{1}{x},$$

which does not possess a Maclaurin's series, can be expanded in ascending powers of $x - 1$ to obtain the Taylor's series

$$\frac{1}{x} = 1 - (x - 1) + (x - 1)^2 - (x - 1)^3 + \dots , \qquad 0 < x < 2.$$

(This series can be obtained directly from equation (13.1) by replacing x by $1 - x$.)

It is not always necessary to use equation (13.2) to obtain the Maclaurin's series for a given function. The following short cuts are available.

(i) A power series may be differentiated or integrated term by term for values of x within its interval of convergence (see Example 13.3 where a series for $\cos x$ is obtained by differentiating the series for $\sin x$).

(ii) Two convergent series may be added, subtracted, multiplied or divided. When dividing series, care must be taken to ensure that the denominator is not zero. (For example the series for $\sin x$ and $\cos x$ are convergent for all x but the series for

$$\tan x = \frac{\sin x}{\cos x}$$

is convergent for $-\pi/2 < x < \pi/2$.)

(iii) A new series may be obtained by simple replacement of the independent variable (see Example 13.3 where the series for $\cos(x^2)$ is obtained from the series for $\cos x$ by replacing x by x^2).

Example 13.2

Use Maclaurin's theorem to obtain the series

$$\sin x = x - \frac{x^3}{3!} + \frac{x^5}{5!} - \frac{x^7}{7!} + \dots .$$

Let

$$f(x) = \sin x,$$

i.e.

$$f(0) = 0.$$

Therefore,

$$\begin{aligned}
f'(x) &= \cos x && \text{and} && f'(0) = 1,\\
f''(x) &= -\sin x && \text{and} && f''(0) = 0,\\
f'''(x) &= -\cos x && \text{and} && f'''(0) = -1,\\
f^{(4)}(x) &= \sin x && \text{and} && f^{(4)}(0) = 0,\\
f^{(5)}(x) &= \cos x && \text{and} && f^{(5)}(0) = 1,\\
f^{(6)}(x) &= -\sin x && \text{and} && f^{(6)}(0) = 0,\\
f^{(7)}(x) &= -\cos x && \text{and} && f^{(7)}(0) = -1,
\end{aligned}$$

etc.

Substituting these values into equation (13.2), we obtain

$$\sin x = x - \frac{x^3}{3!} + \frac{x^5}{5!} - \frac{x^7}{7!} + \dots .$$

We have already seen in Example 13.1 (ii) that this series is convergent for all x.

Example 13.3

By differentiation of the series for $\sin x$, obtain a series for $\cos x$ in ascending powers of x. Hence evaluate

$$\int_0^1 \cos(x^2)\,dx$$

correct to four decimal places.

Differentiating the series obtained in Example 13.2, we obtain

$$\cos x = 1 - \frac{3x^2}{3!} + \frac{5x^4}{5!} - \frac{7x^6}{7!} + \ldots,$$

i.e.

$$\cos x = 1 - \frac{x^2}{2!} + \frac{x^4}{4!} - \frac{x^6}{6!} + \ldots,$$

which is also convergent for all x.

The function

$$\int \cos(x^2)\,\mathrm{d}x$$

cannot be expressed in terms of elementary functions. However, the definite integral

$$\int_a^b \cos(x^2)\,\mathrm{d}x$$

can be estimated using the numerical procedures in Chapter 12. An alternative approach is to express the function $\cos(x^2)$ as an infinite series in ascending powers of x and to integrate this series term by term. For this technique to be successful, it is necessary that the limits of integration are values within the interval of convergence of the infinite series.

The series for $\cos(x^2)$ may be obtained from the series for $\cos x$ by simple replacement of x by x^2, i.e.

$$\cos(x^2) = 1 - \frac{x^4}{2!} + \frac{x^8}{4!} - \frac{x^{12}}{6!} + \ldots,$$

which is convergent for all x.

Therefore,

$$\begin{aligned}\int_0^1 \cos(x^2)\,\mathrm{d}x &= \int_0^1 \left(1 - \frac{x^4}{2} + \frac{x^8}{24} - \frac{x^{12}}{720} + \frac{x^{16}}{40\,320} - \ldots\right)\mathrm{d}x \\ &= \left[x - \frac{x^5}{10} + \frac{x^9}{216} - \frac{x^{13}}{9360} + \frac{x^{17}}{685\,440} - \ldots\right]_0^1 \\ &\approx 1 - 0.1 + 0.004\,630 - 0.000\,107 + 0.000\,001 \\ &= 0.9045 \text{ correct to four decimal places.}\end{aligned}$$

(As a 'rule of thumb', two additional decimal places have been used in the working. The infinite series has been terminated after five terms, subsequent terms being negligible to the required accuracy.)

Example 13.4

(i) Use Maclaurin's theorem to express the function $(1+x)^n$ as an infinite series in ascending powers of x.

(ii) Write down the first three terms of the power series which represents the function $\sqrt{8+x^3}$ and state its interval of convergence.

(iii) Evaluate

$$\int_0^{0.5} \sqrt{8+x^3}\,dx$$

correct to four decimal places.

(iv) Discuss the difficulties associated with the evaluation of

$$\int_2^{2.1} \sqrt{8+x^3}\,dx.$$

(i) Let

$$f(x) = (1+x)^n,$$

i.e.

$$f(0) = 1.$$

Therefore,

$$f'(x) = n(1+x)^{n-1} \quad \text{and} \quad f'(0) = n,$$

$$f''(x) = n(n-1)(1+x)^{n-2} \quad \text{and} \quad f''(0) = n(n-1),$$

$$f'''(x) = n(n-1)(n-2)(1+x)^{n-3} \quad \text{and} \quad f'''(0) = n(n-1)(n-2),$$

etc.

Substituting these numbers into equation (13.2), we obtain

$$(1+x)^n = 1 + nx + \frac{n(n-1)}{2!}x^2 + \frac{n(n-1)(n-2)}{3!}x^3 + \ldots \tag{13.4}$$

If n is a positive integer, the series terminates after $n+1$ terms and the result is valid for *all* x. (For example, $(1+x)^2 = 1 + 2x + x^2$ for all x.) However, if n is *not* a positive integer, the series is infinite with interval of convergence $-1 < x < 1$. Equation (13.4) is known as the **binomial series.**

(ii)

$$\sqrt{8+x^3} = \sqrt{8}\left(1 + \frac{x^3}{8}\right)^{1/2}.$$

We apply equation (13.4) with $n = \frac{1}{2}$ and with x replaced by $x^3/8$;

$$\sqrt{8+x^3} = \sqrt{8}\left\{1 + \frac{1}{2}\frac{x^3}{8} + \frac{\frac{1}{2}(-\frac{1}{2})}{2!}\left(\frac{x^3}{8}\right)^2 + \ldots\right\}$$

$$= \sqrt{8}\left(1 + \frac{x^3}{16} - \frac{x^6}{512} + \ldots\right),$$

which is convergent for $-1 < x^3/8 < 1$, i.e. the interval of convergence is $-2 < x < 2$.

(iii) $\int \sqrt{8+x^3}\,\mathrm{d}x$ is another integral which cannot be expressed in terms of elementary functions. The *definite* integral

$$\int_0^{0.5} \sqrt{8+x^3}\,\mathrm{d}x$$

may be evaluated approximately using the infinite series derived in (ii) (the limits of integration 0 and 0.5 lie within the interval of convergence $-2 < x < 2$).

$$\int_0^{0.5} \sqrt{8+x^3}\,\mathrm{d}x = \sqrt{8}\int_0^{0.5}\left(1 + \frac{x^3}{16} - \frac{x^6}{512} + \ldots\right)\mathrm{d}x$$

$$= \sqrt{8}\left[x + \frac{x^4}{64} - \frac{x^7}{3584} + \ldots\right]_0^{0.5}$$

$$= \sqrt{8}(0.5 + 0.000\,976\,6 - 0.000\,002\,2 + \ldots)$$

$= 1.4170$ correct to four decimal places.

(iv) The series for $\sqrt{8+x^3}$ derived in (ii) may not be used to calculate

$$\int_2^{2.1} \sqrt{8+x^3}\,\mathrm{d}x$$

since the limits of integration do not lie in the interval of convergence $-2 < x < 2$. However, we may use equation (13.3) to express $\sqrt{8+x^3}$ as an infinite series in ascending powers of $x - 2$. Such a series will be convergent for values of x near to $x = 2$ and may be used for the evaluation of

$$\int_2^{2.1} \sqrt{8+x^3}\,\mathrm{d}x.$$

It is left as an exercise for the reader to show that

$$\sqrt{8+x^3} = 4 + \tfrac{3}{2}(x-2) + \tfrac{15}{32}(x-2)^2 - \tfrac{13}{256}(x-2)^3 + \ldots,$$

and

$$\int_2^{2.1} \sqrt{8+x^3}\,\mathrm{d}x = 0.4077$$

correct to four decimal places.

Problems

1. Use Maclaurin's theorem to derive the power-series expansions given in problem 1 of section 13.2 for the functions $\ln(1+x)$, e^x, $\sinh x$. Deduce that

$$\cosh x = 1 + \frac{x^2}{2!} + \frac{x^4}{4!} + \frac{x^6}{6!} + \ldots,$$

for all x.

2. Use the binomial series to obtain a series for

$$\frac{1}{\sqrt{1-x^2}}$$

and state the interval of convergence. By integration, deduce a series for $\sin^{-1} x$ and use it to evaluate

$$\int_0^{0.1} \sin^{-1} x \, dx$$

correct to six decimal places. Check your answer using integration by parts.

3. Write down a series for $\sqrt{1+x^3}$ and use it to calculate

$$\int_0^{0.1} \sqrt{1+x^3} \, dx$$

correct to three decimal places.

4. Use the series for $\ln(1+x)$ derived in problem 1 to obtain series for the functions

$$\ln(1-x)$$

and

$$\ln\left(\frac{1+x}{1-x}\right).$$

Choosing a suitable value of x within the interval of convergence of one of these series, calculate $\ln(\frac{11}{9})$ correct to five decimal places.

5. Evaluate

$$\int_0^1 \sin(x^2) \, dx$$

correct to four decimal places.

6. Derive the series for $\cos x$ in ascending powers of x by *integrating* the series for $\sin x$. Divide the series for $\sin x$ by the series for $\cos x$ to obtain

$$\tan x = x + \frac{x^3}{3} + \frac{2x^5}{15} + \ldots, \qquad -\frac{\pi}{2} < x < \frac{\pi}{2}.$$

7. Obtain the following infinite series for e^{-x^2}.
   (i) In ascending powers of x as far as the term in x^6.
   (ii) In ascending powers of $x - 1$ as far as the term in $(x - 1)^4$.
   Use the series to evaluate

$$\int_0^{0.1} e^{-x^2}\, dx$$

and

$$\int_1^{1.1} e^{1-x^2}\, dx$$

correct to five decimal places.

8. Write down the series in ascending powers of x which represents the function

$$\frac{1}{1 + x^2}.$$

   (i) Deduce that

$$\tan^{-1} x = x - \frac{x^3}{3} + \frac{x^5}{5} - \frac{x^7}{7} + \ldots, \qquad -1 < x < 1.$$

   (ii) By multiplication of two infinite series, obtain the Maclaurin's series for the function

$$\frac{e^x}{1 + x^2}$$

as far as the term in x^4. State the interval of convergence.

Answers

2. 0.005 004.

3. $1 + \frac{x^3}{2} - \frac{x^6}{8} + \frac{x^9}{16} - \ldots, \; -1 < x < 1,$

   0.100 01.

4. $\ln(1 - x) = -x - \frac{x^2}{2} - \frac{x^3}{3} - \frac{x^4}{4} - \ldots, \; -1 < x < 1,$

$$\ln\left(\frac{1+x}{1-x}\right) = 2\left(x + \frac{x^3}{3} + \frac{x^5}{5} + \ldots\right), \quad -1 < x < 1,$$

0.200 67.

5. 0.3103.

7. (i) $1 - x^2 + \dfrac{x^4}{2!} - \dfrac{x^6}{3!} + \ldots$, all x.

   (ii) $e^{-1}\{1 - 2(x-1) + (x-1)^2 + \frac{2}{3}(x-1)^3 - \frac{5}{6}(x-1)^4 + \ldots\}$, all x,

   0.099 67, 0.090 35.

8. $1 - x^2 + x^4 - x^6 + \ldots, \quad -1 < x < 1.$

   (ii) $1 + x - \dfrac{x^2}{2} - \dfrac{5x^3}{6} + \dfrac{13x^4}{24} + \ldots, \quad -1 < x < 1.$

Further Problems for Chapter 13

1. Determine the numerical values of percentage errors made when the approximation

$$\sin x = x - \frac{x^3}{6}$$

   is used to calculate sin 1 and sin(1.5).

2. Use Theorem 13.1 to show that the following series is convergent:

$$1 + 2 \times 0.1 + 3 \times (0.1)^2 + 4 \times (0.1)^3 + 5 \times (0.1)^4 + \ldots .$$

   Is the series still convergent if the first two terms are replaced by $1500 + 20\,000 \times 0.1$.?

3. Write down the radius of convergence of the power series

$$1 + 2x + 4x^2 + 8x^3 + 16x^4 + \ldots .$$

4. Given that

$$S(n) = \sum_{r=1}^{n} u_r$$

   show that $u_n = S(n) - S(n-1)$. If the infinite series

$$\sum_{r=1}^{\infty} u_r$$

   is convergent, deduce that $\lim_{r\to\infty} u_r = 0$. (Note that the converse is not true; $\lim_{r\to\infty} u_r$ may be zero for a divergent series.)

5. Use the series for e^x to calculate e correct to six decimal places.

6. Write down the series in ascending powers of x for the function

$$\frac{1}{1-x^2}.$$

Integrate this series to show that

$$\tanh^{-1} x = x + \frac{x^3}{3} + \frac{x^5}{5} + \ldots, \qquad -1 < x < 1.$$

Compare with the series for

$$\ln\left(\frac{1+x}{1-x}\right)$$

(problem 4, section 13.3).

7. Explain why the function $f(x) = x^{5/2}$ does not possess a Maclaurin's series.

8. State the error in the following:

$$\int_0^2 \sqrt{1+x^5}\, dx = \int_0^2 \left(1 + \frac{x^5}{2} - \frac{x^{10}}{8} + \ldots\right) dx.$$

9. Write down a third-degree polynomial approximation for the function $\cos(x\sqrt{x})$ which can be used when x is small.

10. Use the approximation in the previous question to estimate the value of

$$\int_0^{0.5} \cos(x\sqrt{x})\, dx.$$

Check your answers by applying Simpson's rule with two strips to the above integral.

11. Given that $i = \sqrt{-1}$, i.e. $i^2 = -1$, $i^3 = -i$, $i^4 = 1, \ldots$, etc., use the series for e^x to obtain the series for e^{ix} and show that

$$e^{ix} = \cos x + i \sin x.$$

Answers

1. 1% and 6.4% approximately.
2. Yes.
3. 0.5.
5. 2.718 282.

6. $1 + x^2 + x^4 + x^6 + \dots, \quad -1 < x < 1.$
7. $f'''(x)$ and higher derivatives do not exist when $x = 0$.
8. $1 + \dfrac{x^5}{2} - \dfrac{x^{10}}{8} + \dots$ is not convergent for $x = 2$.
9. $1 - \dfrac{x^3}{2}$.
10. 0.492.

14

Differential Equations

14.1 INTRODUCTION

Differential equations occur throughout the branches of engineering and science, particularly in dynamics and other situations where the rates of change in a quantity are related to the quantity and to functions of the independent variable. A differential equation is an equation which includes a combination of the derivatives of the unknown function. If only the first derivative is present, the equation is said to be a *first-order* differential equation. We have already met such an equation in Chapter 8, i.e. $\mathrm{d}y/\mathrm{d}t = ky$, in the study of growth and decay problems. In section 4.5 we studied equations of the type $\mathrm{d}^2s/\mathrm{d}t^2 = f(t)$ which is a *second-order* differential equation where s is the unknown function and $f(t)$ is given. A differential equation of order n includes the nth derivative of the unknown function and may also include *lower*-order derivatives. The *solution* of a differential equation is an expression of the unknown function in terms of the independent variable. If the differential equation has order n, then n integrations are required to obtain the solution (although some of the solution techniques tend to hide this important feature). Consequently the *general solution* of a differential equation of order n will include n arbitrary constants, one for each integration. For example, the first-order equation $\mathrm{d}y/\mathrm{d}t = ky$ has the general solution $y = Ce^{kt}$, where C is an arbitrary constant. The general solution of the second-order differential equation $\mathrm{d}^2s/\mathrm{d}t^2 = 10$ can be obtained by two simple integrations, i.e. $s = 5t^2 + At + B$, where A and B are the two arbitrary constants.

Any real problem in science and engineering will not of course have an arbitrary solution. Additional information in the form of *initial conditions* or *boundary conditions* (already defined in section 4.5) is required so that the arbitrary constants in the general solution may be evaluated. If the differential equation is of order n, then n such conditions must be specified. For example, returning to the first-order differential equation $\mathrm{d}y/\mathrm{d}t = ky$ with general solution $y = C\,\mathrm{e}^{kt}$, the single condition $y = y_0$ when $t = 0$ implies that $C = y_0$ and gives the solution $y = y_0\,\mathrm{e}^{kt}$. Similarly the second-order diferential equation $\mathrm{d}^2s/\mathrm{d}t^2 = 10$ with general solution $s = 5t^2 + At + B$ requires two conditions to be specified. Suppose that these conditions are $s = 0$ when $t = 0$ and $s = 20$ when $t = 1$. The first condition requires that $B = 0$, and $A = 15$ is necessary to satisfy the second condition. Therefore the solution is $s = 5t^2 + 15t$ (the reader should look again at Example 4.9 where a very similar problem was considered).

The technique of solution applied above to the equation $\mathrm{d}^2s/\mathrm{d}t^2 = 10$ is known as *direct integration*. This method can always be used for equations of the form $\mathrm{d}^ny/\mathrm{d}x^n = f(x)$ where n integrations will express y in terms of x and n arbitrary constants. Most equations occurring in real problems are not of this type and special methods of solution are required. Some of these techniques are considered in sections 14.3 and 14.4 for first- and second-order differential equations. In section 14.2 a number of examples are considered of situations where differential equations occur in engineering and science.

14.2 OCCURRENCE OF DIFFERENTIAL EQUATIONS IN PRACTICAL SITUATIONS

Since special solution techniques have not yet been studied, the solution is *given* in most of the examples and problems considered in this section. The given solution is then *verified* by substitution into the differential equation.

Example 14.1

An electric circuit consists of a pure resistance of $R\ \Omega$ and a coil of inductance L H connected in series to an alternating current supply whose voltage is given by $V = E \sin wt$, where E and w are constants. The current i A in the circuit satisfies the differential equation

$$L\frac{\mathrm{d}i}{\mathrm{d}t} + Ri = E \sin wt$$

and the condition $i = 0$ when $t = 0$. Show that

$$i = \frac{E}{R^2 + L^2w^2}\,(R \sin wt - Lw \cos wt + Lw\,\mathrm{e}^{-Rt/L}).$$

The solution is *verified* as follows.

(i) By showing that it satisfies the condition $i = 0$ when $t = 0$ (this is clearly true).

(ii) By showing that it satisfies the differential equation. We differentiate i, and substitute i and $\mathrm{d}i/\mathrm{d}t$ into the given differential equation.

$$\frac{\mathrm{d}i}{\mathrm{d}t} = \frac{E}{R^2 + L^2w^2}\,(Rw \cos wt + Lw^2 \sin wt - Rw\,\mathrm{e}^{-Rt/L}).$$

Therefore,

$$L\frac{di}{dt}+Ri=\frac{E}{R^2+L^2w^2}\{(L^2w^2+R^2)\sin wt$$

$$+(LRw-RLw)\cos wt-(LRw-RLw)\,e^{-Rt/L}\}$$

$$=E\sin wt,$$

and the given differential equation is satisfied.

(A method for solving this type of first-order differential equation is given in section 14.3.)

Example 14.2

In a bimolecular chemical reaction, substances A and B form molecules of a substance C. If a and b are the original concentrations of A and B respectively and x is the concentration of C at a given instant, then the following differential equation may be established to describe the reaction:

$$\frac{dx}{dt}=k(a-x)(b-x),$$

where k is a constant.

Show that

$$x=\frac{ab\{1-e^{(a-b)kt}\}}{b-a\,e^{(a-b)kt}},\quad a\neq b.$$

Firstly, x satisfies the condition that $x=0$ when $t=0$ (no substance C present at the start of the reaction).

To *verify* the given solution, it is necessary to differentiate x with respect to t:

$$\frac{dx}{dt}=\frac{(b-aE)\,ab\,\{-k(a-b)E\}-ab(1-E)\,\{-ka(a-b)E\}}{(b-aE)^2}$$

$$=\frac{kabE(b-a)^2}{(b-aE)^2},$$

where E denotes $e^{(a-b)kt}$. Also,

$$k(a-x)(b-x)=k\,\frac{\{a(b-aE)-ab(1-E)\}\{b(b-aE)-ab(1-E)\}}{(b-aE)^2}$$

$$=\frac{kabE(b-a)^2}{(b-aE)^2}.$$

Therefore,

$$\frac{dx}{dt} = k(a-x)(b-x)$$

and the differential equation is satisfied by the given solution.

(The special technique for solving a differential equation of this type is given in section 14.3.)

Example 14.3

The differential equation

$$m\frac{d^2x}{dt^2} + C\frac{dx}{dt} + kx = k$$

is a crude model used to describe the response of a car suspension system to a sudden bump in the road. m, k and C are positive constants related to the mass of the car, the stiffness of the springs and the efficiency of the shock absorbers. Given that both x and dx/dt are zero at the time $t = 0$, when the car hits the bump, show that the following are true.

(i) For the case $C = 0$ (shock absorber failure), $x = 1 - \cos wt$, $w = \sqrt{k/m}$.

(ii) For the case $C^2/4m^2 = k/m = w^2$ (called critical damping), $x = 1 - (1 + wt)\,e^{-wt}$.

For both cases, we *verify* that the given solution satisfies the differential equation and the initial conditions.

(i) $x = 1 - \cos wt$.

$$\frac{dx}{dt} = w \sin wt.$$

$$\frac{d^2x}{dt^2} = w^2 \cos wt.$$

Clearly, $x = dx/dt = 0$ when $t = 0$.

$$m\frac{d^2x}{dt^2} + kx = mw^2 \cos wt + k(1 - \cos wt)$$

$$= k \cos wt + k - k \cos wt,$$

since $w^2 = k/m$. Therefore,

$$m\frac{d^2x}{dt^2} + kx = k,$$

and the differential equation is satisfied.

(ii) $x = 1 - (1 + wt)\,e^{-wt}$.

$$\frac{dx}{dt} = w(1 + wt)\,e^{-wt} - w\,e^{-wt}$$

$$= w^2 t\,e^{-wt}.$$

$$\frac{d^2x}{dt^2} = w^2(1 - wt)\,e^{-wt}.$$

Clearly, $x = dx/dt = 0$ when $t = 0$.

$$m\frac{d^2x}{dt^2} + C\frac{dx}{dt} + kx = mw^2(1 - wt)\,e^{-wt} + Cw^2 t\,e^{-wt}$$

$$+ k\{1 - (1 + wt)\,e^{-wt}\}$$

$$= (mw^2 - k)\,e^{-wt} + w^2\left(C - wm - \frac{k}{w}\right) t\,e^{-wt} + k$$

$$= k,$$

since $mw^2 = k$ and $wm + k/w = 2wm = C$. Therefore the differential equation is satisfied.

(A special technique for solving this type of second-order differential equation is considered in section 14.4.)

Example 14.4

A suspension bridge is supported by a cable which hanges between the tops of two towers of height h, the distance between the towers being l. The lowest point of the cable is above the mid-point of the horizontal line through the base of the towers. If the cable supports a load w per unit horizontal distance the shape of the cable satisfies the differential equation

$$\frac{d^2y}{dx^2} = \frac{w}{H}$$

where H is the tension in the cable at its lowest point and the axes Oxy are defined as shown in Fig. 14.1.

Determine the equation for the shape taken up by the cable.

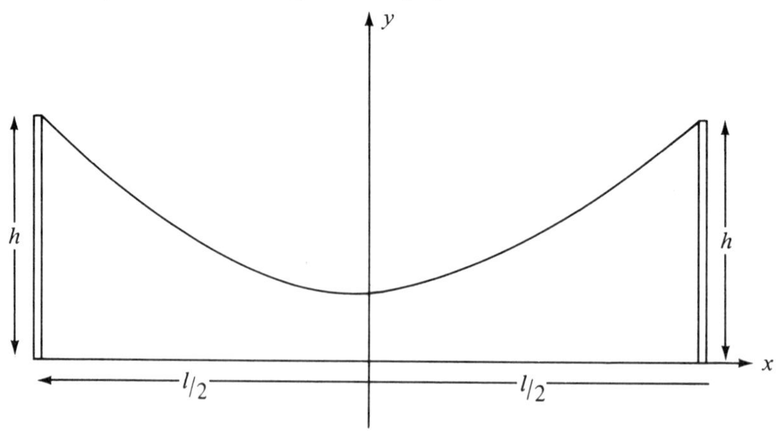

Fig. 14.1

Unlike Examples 14.1–14.3 the solution is not *given*, but no special techniques are required to obtain the solution. Direct integration gives

$$\frac{dy}{dx} = \frac{wx}{H} + A$$

and

$$y = \frac{wx^2}{2H} + Ax + B.$$

The conditions $dy/dx = 0$ when $x = 0$ and $y = h$ when $x = l/2$ must be satisfied. Therefore, $A = 0$ and $B = h - wl^2/8H$, and the equation describing the shape of the cable is

$$y = \frac{w}{8H}(4x^2 - l^2) + h.$$

Problems

1. A model for a population, which is restricted to some maximum number M, is based on the assumption that the rate of growth is proportional to the product of the number X in the population and $M - X$, i.e. $dX/dt = kX(M - X)$, where k is a constant. Show that

$$X = \frac{M}{1 + e^{-kM(t-t_1)}}$$

where $t = t_1$ is the time at which $X = M/2$.

2. A model of species competing for their food supply is given by the simultaneous differential equations

$$\frac{dx}{dt} = ax - by$$

and

$$\frac{dy}{dt} = cy - dx,$$

where a, b, c and d are all positive constants and x and y represent the scaled population sizes.

(i) Differentiate the first equation and eliminate y to show that x satisfies the second-order differential equation

$$\frac{d^2x}{dt^2} - (a + c)\frac{dx}{dt} + (ac - db)x = 0.$$

(ii) Given that $a = c = 2$, $b = d = 1$, $x = 100$ and $y = 200$ when $t = 0$, show that the initial value of $\mathrm{d}x/\mathrm{d}t$ is zero and that $x = 150\,\mathrm{e}^{t} - 50\,\mathrm{e}^{3t}$.

(iii) Express y in terms of t and determine the time when one of the species is eliminated.

3. Suppose that X g of salt are required to produce saturation in M g of water. If x g is the amount of salt left undissolved t s after X g have been placed in M g of water, then x satisfies the differential equation

$$\frac{\mathrm{d}x}{\mathrm{d}t} = -\frac{kx^2}{M},$$

where k is a constant. Show that

$$\frac{1}{x} - \frac{1}{X} = \frac{kt}{M}.$$

4. A parachutist falls freely, starting from rest at an altitude of 1000 m. Calculate the parachutist's velocity and altitude after 4 s (neglect air resistance and take $g = 9.81\ \mathrm{m/s^2}$). The parachute opens at this time, creating a drag force equal to Kv where v is the velocity of the parachutist. The motion is modelled approximately by the differential equation

$$m\frac{\mathrm{d}v}{\mathrm{d}t} = mg - Kv,$$

where the velocity is positive for descending motion, and m is the mass of the parachutist plus parachute. Show that for $m = 100$ kg and $K = 120\ \mathrm{N\,s\,m^{-1}}$ the solution of the differential equation is

$$v = 8.175 + 3774.6\,\mathrm{e}^{-1.2t}, \qquad t \geqslant 4,$$

where $t = 4$ corresponds to the time when the parachute was opened. Estimate the velocity of the parachutist at sea level.

5. A heavy beam of length $2l$ is simply supported at both ends as shown in Fig. 14.2. With respect to the axes shown in the figure the shape taken up by the beam satisfies the equation

$$\frac{wx(l - 0.5x)}{I} = \frac{E}{\rho},$$

where w is the load per unit length, I is the second moment of area of the beam section, E is Young's modulus for the beam and ρ is the radius of curvature at the point (x, y) on the beam. Making the approximation

$$\rho = \frac{1}{(\mathrm{d}^2y/\mathrm{d}x^2)},$$

determine the equation for the shape of the beam. Show that the maximum deflection of the beam is

$$\frac{5wl^4}{24EI}.$$

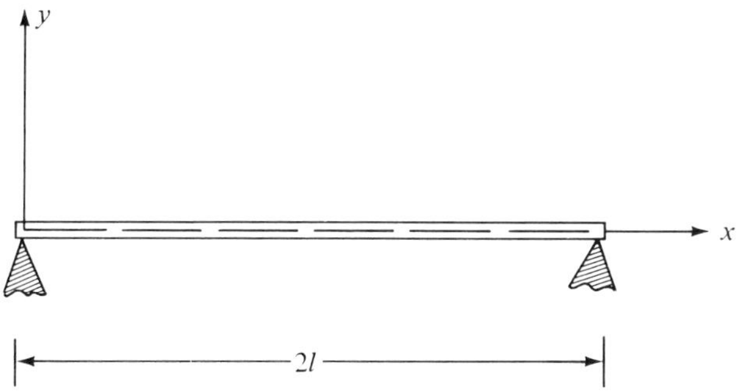

Fig. 14.2

Answers

2. (iii) $y = 150\,e^t + 50\,e^{3t}$, $\frac{1}{2}\ln 3$ (units of time).
4. 39.24 m/s, 921.52 m; 8.175 m/s.
5. $y = \dfrac{w}{24EI}(4lx^3 - x^4 - 8l^3x)$.

14.3 FIRST-ORDER DIFFERENTIAL EQUATIONS

The general first-order differential equation takes the form $dy/dx = F(x, y)$, where $F(x, y)$ is a function of the *two* variables x and y. If $F(x, y)$ is a function of x only, then direct integration may be applied to obtain a solution. Otherwise, special techniques are required and, in this section, we consider two such methods.

The first method applies when $F(x, y)$ can be factorized into two factors, one of which is a function of x only and the other a function of y only. It is then possible to **separate the variables** and to complete the integration. For example, $dy/dx = 2y^2x + y^2$ can be written $dy/dx = y^2(2x + 1)$ and, applying the method of separation of variables, we obtain

$$\frac{dy}{y^2} = (2x + 1)\,dx.$$

We now apply the operation of integration to both sides of the above equation, i.e. $-1/y = x^2 + x + C$ (only *one* constant required) and the solution is

$$y = -\frac{1}{x^2 + x + C}.$$

The second method applies only to *linear* first-order equations which take the form

$$\frac{dy}{dx} = Q(x) - yP(x),$$

i.e.

$$\frac{dy}{dx} + P(x)y = Q(x) \tag{14.1}$$

where $P(x)$ and $Q(x)$ are functions of x only.

Equation (14.1) may be simplified by multiplying both sides by $\exp\left(\int P(x)\,dx\right)$ (called an **integrating factor**), i.e.

$$\exp\left(\int P\,dx\right)\frac{dy}{dx} + P\exp\left(\int P\,dx\right)y = \exp\left(\int P\,dx\right)Q \tag{14.2}$$

(where $P(x)$ and $Q(x)$ have been abbreviated to P and Q).

The left hand side of equation (14.2) can be written as $(d/dx)\left\{\exp\left(\int P\,dx\right)y\right\}$ since $P\exp\left(\int P\,dx\right)$ is the derivative of $\exp\left(\int P\,dx\right)$, i.e. equation (14.2) becomes

$$\frac{d}{dx}\left\{\exp\left(\int P\,dx\right)y\right\} = \exp\left(\int P\,dx\right)Q,$$

which can be solved by direct integration since the right-hand side is a function of x only. For example, the equation $dy/dx + 2y = e^{-3x}$ is linear with $P(x) = 2$. The integrating factor is $\exp\left(\int 2\,dx\right) = e^{2x}$ and, multiplying the differential equation by this factor, we obtain

$$e^{2x}\frac{dy}{dx} + 2\,e^{2x}y = e^{-x}.$$

The left-hand side of this equation is $(d/dx)\,(e^{2x}y)$ (check, using the product rule), and the equation becomes

$$\frac{d}{dx}(e^{2x}y) = e^{-x}.$$

One integration gives

$$e^{2x}y = \int e^{-x}\,dx = -e^{-x} + C,$$

and the general solution is

$$y = -e^{-3x} + C\,e^{-2x},$$

where C is an arbitrary constant.

Example 14.5
Solve the following differential equations.

(i) $$x\frac{dy}{dx} - y = y^3$$

given that $y = 1$ when $x = 1$.

(ii) $$x\frac{dy}{dx} - 3y = x^5$$

given that $y = 0$ when $x = 2$.

(i) The differential equation

$$x\frac{dy}{dx} - y = y^3$$

is not linear in y because of the presence of the term y^3. However, it can be written in the form

$$\frac{dy}{dx} = \frac{y^3 + y}{x} .$$

Separating the variables and integrating,

$$\int \frac{dy}{y^3 + y} = \int \frac{dx}{x} .$$

The integral on the left-hand side is carried out using the method of partial fractions:

$$\frac{1}{y^3 + y} = \frac{1}{y(1 + y^2)} = \frac{1}{y} - \frac{y}{1 + y^2} .$$

Therefore,

$$\int \frac{dy}{y} - \int \frac{y\,dy}{1 + y^2} = \int \frac{dx}{x} ,$$

i.e. $\ln|y| - \frac{1}{2}\ln(1 + y^2) = \ln|x| + C.$
The condition $y = 1$ when $x = 1$ requires that $C = -\frac{1}{2}\ln 2$. Multiplying by 2 and using $2\ln|x| = \ln(x^2)$, we obtain

$$\ln(y^2) - \ln(1 + y^2) = \ln(x^2) - \ln 2,$$

i.e.

$$\ln\left(\frac{y^2}{1 + y^2}\right) = \ln\left(\frac{x^2}{2}\right),$$

and

$$\frac{y^2}{1+y^2} = \frac{x^2}{2}$$

is the solution of the differential equation in implicit function form.

(ii) The differential equation

$$x\frac{dy}{dx} - 3y = x^5$$

is linear in y but, before applying the integrating factor technique, it must be expressed in the standard form of equation (14.1), i.e.

$$\frac{dy}{dx} + \left(\frac{-3}{x}\right)y = x^4. \qquad (14.3)$$

Therefore $P(x) = -3/x$ and the integrating factor is given by

$$\exp\left(\int \frac{-3}{x}\,dx\right) = e^{-3\ln x}$$

$$= e^{\ln(x^{-3})}$$

$$= \frac{1}{x^3},$$

by definition of a logarithm. We multiply equation (14.3) by the integrating factor to obtain

$$\frac{1}{x^3}\frac{dy}{dx} + \left(\frac{-3}{x^4}\right)y = x. \qquad (14.4)$$

(This step is not strictly necessary since the next stage can be written down directly knowing that the left-hand side will be (d/dx) (integrating factor $\times$ y). However, if we use the product rule to carry out the differentiation on the left-hand side of equation (14.5) we can check it by comparison with equation (14.4).)

$$\frac{d}{dx}\left(\frac{1}{x^3}y\right) = x. \qquad (14.5)$$

Integrating equation (14.5),

$$\frac{1}{x^3}y = \frac{x^2}{2} + C.$$

The condition $y = 0$ when $x = 2$ requires that $C = -2$, i.e.

$$y = \frac{x^5}{2} - 2x^3$$

is the solution of the linear differential equation.

Example 14.6

Solve the differential equation given in Example 14.2 using the separation of variable technique.

$$\frac{\mathrm{d}x}{\mathrm{d}t} = k(a-x)(b-x), \quad x = 0 \quad \text{when } t = 0.$$

Separating the variables and integrating,

$$\int \frac{\mathrm{d}x}{(a-x)(b-x)} = \int k \,\mathrm{d}t,$$

i.e.

$$\frac{1}{a-b}\left(\int \frac{-\mathrm{d}x}{a-x} + \int \frac{\mathrm{d}x}{b-x}\right) = \int k \,\mathrm{d}t$$

and

$$\ln(a-x) - \ln(b-x) = k(a-b)t + C.$$

To satisfy $x = 0$ when $t = 0$ requires $C = \ln a - \ln b$. Therefore,

$$\ln\left\{\frac{b(a-x)}{a(b-x)}\right\} = k(a-b)t,$$

i.e.

$$\frac{b(a-x)}{a(b-x)} = \mathrm{e}^{k(a-b)t},$$

which can be solved for x to obtain the solution given in Example 14.2.

Problems

1. Solve the following first-order differential equations giving the general solution unless a condition is specified.

   (i) $\dfrac{\mathrm{d}y}{\mathrm{d}x} + \dfrac{y}{x} = x^2 - 1.$

   (ii) $(1 + x^2)\dfrac{\mathrm{d}y}{\mathrm{d}x} + 4xy = 0, \quad y = 1$ when $x = 1$.

   (iii) $\dfrac{\mathrm{d}y}{\mathrm{d}x} - y \tan x = 2 \sin x.$

   (iv) $2t\dfrac{\mathrm{d}x}{\mathrm{d}t} + x^2 = 1.$

(v) $(1-t^2)\dfrac{\mathrm{d}y}{\mathrm{d}t} + 2ty = t - t^3$, $y = 0$ when $t = 0$.

2. Obtain the general solution of the differential equation

$$\frac{\mathrm{d}y}{\mathrm{d}t} + py = q,$$

where p and q are constants. Use your result to write down the solution of the following equations.
(i) The open parachute equation given in problem 4 of section 14.2.
(ii) The equation for the circuit described in Example 14.1 with the alternating voltage supply replaced by a constant voltage $V = E$.

3. Apply the method of separation of variables to obtain the solutions of the differential equations given in problems 1 and 3 in section 14.2.

Answers

1. (i) $y = \dfrac{x^3}{4} - \dfrac{x}{2} + \dfrac{C}{x}$.

(ii) $y = \dfrac{4}{(1+x^2)^2}$.

(iii) $y = C\sec x - \cos x$.
(iv) $1 + x = Ct(1 - x)$.

(v) $y = (1-t^2)\ln\left(\dfrac{1}{\sqrt{1-t^2}}\right)$

2. $y = \dfrac{q}{p} + C\,\mathrm{e}^{-pt}$.

(ii) $i = \dfrac{E}{R}(1 - \mathrm{e}^{-Rt/L})$.

14.4 LINEAR SECOND-ORDER DIFFERENTIAL EQUATIONS WITH CONSTANT COEFFICIENTS

In this section, we consider equations of the type

$$\frac{\mathrm{d}^2 y}{\mathrm{d}t^2} + p\frac{\mathrm{d}y}{\mathrm{d}t} + qy = f(t), \tag{14.6}$$

where p and q are constants.

Many situations in engineering and science are modelled by equations of this type, particularly in dynamics where they arise in the application of Newton's second law of motion. A detailed study of equation (14.6) is important since many higher-order and non-linear engineering systems are approximated by second-order linear equations. For example the differential equation model of the car suspension system described in Example 14.3 is a crude, but useful, approximation to the real-life situation.

Closely associated with equation (14.6) is the *reduced* (or *homogeneous*) equation

$$\frac{d^2y}{dt^2} + p\frac{dy}{dt} + qy = 0. \tag{14.7}$$

The solution techniques described in this section require us to solve equation (14.7) before the solution of equation (14.6) can be determined. The following theorems are necessary to establish the method of solution. An indication of the proof is given in each case.

Theorem 14.1

If $y = y_1(t)$ and $y = y_2(t)$ are solutions of equation (14.7) which are *independent* (i.e. $y_1(t)$ is not a constant multiple of $y_2(t)$), then the general solution of equation (14.7) is

$$y = Ay_1(t) + By_2(t), \tag{14.8}$$

where A and B are arbitrary constants.

Proof

Substitution for y from equation (14.8) into equation (14.7) verifies that $y = Ay_1(t) + By_2(t)$ is a solution of equation (14.7). Since $y_1(t)$ and $y_2(t)$ are independent the solution contains the two arbitrary constants required for the general solution of a second-order differential equation.

Theorem 14.2 (The solution of equation (14.7))

If $s = s_1$ and $s = s_2$ are the solutions of the quadratic equation $s^2 + ps + q + 0$, then the general solution of

$$\frac{d^2y}{dt^2} + p\frac{dy}{dt} + qy = 0 \tag{14.7}$$

is obtained from one of the following cases.

(i) If s_1 and s_2 are *real* and $s_1 \neq s_2$, then

$$y = A\,e^{s_1 t} + B\,e^{s_2 t}. \tag{14.9}$$

(ii) If s_1 and s_2 are *real* and $s_1 = s_2$, then

$$y = (A + Bt)\,e^{s_1 t}. \tag{14.10}$$

(iii) If s_1 and s_2 are *complex conjugates*, i.e. of the form $u \pm iv$ where $i = \sqrt{-1}$ and u and v are real constants, then

$$y = e^{ut}\,(A \cos vt + B \sin vt). \tag{14.11}$$

(In each case, A and B are arbitrary constants.)

Proof

(i) Since $s_1 \neq s_2$, $e^{s_1 t}$ and $e^{s_2 t}$ are independent functions. We can show by substitution into equation (14.7) that $y = e^{s_1 t}$ and $y = e^{s_2 t}$ *are* solutions, and the result follows from Theorem 14.1.

(ii) $y = e^{s_1 t}$ and $y = t\,e^{s_2 t}$ are independent and satisfy equation (14.7), and the result follows from Theorem 14.1.

(iii) From (i), we would expect the general solution to be

$$y = C_1\, e^{(u+iv)t} + C_2\, e^{(u-iv)t}$$

where C_1 and C_2 are arbitrary constants. Using the result from the further problem 11 in Chapter 13 we obtain

$$y = e^{ut}\left\{C_1(\cos vt + i \sin vt) + C_2(\cos vt - i \sin vt)\right\}.$$

We define $A = C_1 + C_2$ and $B = i(C_1 - C_2)$, (A and B are therefore arbitrary) to give the required form $y = e^{ut}(A \cos vt + B \sin vt)$.

Theorem 14.3 (The solution of equation (14.6))

If $y = y_{\mathrm{CF}}$ is the *general* solution of

$$\frac{d^2 y}{dt^2} + p\frac{dy}{dt} + qy = 0 \tag{14.7}$$

and $y = y_{\mathrm{PI}}$ is *any* solution of

$$\frac{d^2 y}{dt^2} + p\frac{dy}{dt} + qy = f(t), \tag{14.6}$$

then the general solution of equation (14.6) is given by

$$y = y_{\mathrm{CF}} + y_{\mathrm{PI}}. \tag{14.12}$$

Proof

Firstly, we note that the term y_{CF} in equation (14.12) includes the required two arbitrary constants.

When $y = y_{\mathrm{CF}}$ is substituted into the left-hand side of equation (14.6), the result is zero since $y = y_{\mathrm{CF}}$ satisfies equation (14.7). When $y = y_{\mathrm{PI}}$ is substituted into the left-hand side of equation (14.6), $f(t)$ is obtained since $y = y_{\mathrm{PI}}$ is a solution of equation (14.6).

It follows that $y = y_{\mathrm{CF}} + y_{\mathrm{PI}}$ satisfies equation (14.6) and is the general solution.

The function y_{CF} is called the **complementary function** and y_{PI} is called the **particular integral**.

Our method of solution for equation (14.6) is to obtain y_{CF} by application of Theorem 14.2 and to determine y_{PI} by choosing a suitable 'trial solution' related to the function $f(t)$ on the right-hand side of the equation. Suggested trial functions are given in the following table.

$f(t)$	Trial function
A constant	$y = K$
A polynomial in t of degree n	$y = K_1 t^n + K_2 t^{n-1} + \ldots + K_{n+1}$
A multiple of e^{at}	$y = K e^{at}$
A combination of $\sin nt$ and $\cos nt$	$y = K_1 \cos nt + K_2 \sin nt$

The trial function is substituted into equation (14.6) and the constants $K, K_1, K_2, \ldots$ are chosen so that the equation is satisfied.

If the given trial function includes part of the function y_{CF} then the technique fails and it is necessary to multiply the trial function given in the table by t.

The method of solution based on trial functions is demonstrated in Examples (14.7)–(14.11).

More sophisticated methods are available for the solution of linear differential equations with constant coefficients. For example, *inverse operator methods* and *variation of parameters* are summarized in Ayres (1972). *Integral transform techniques* are presented in Johnson (1984).

Example 14.7

Determine the general solution of the following differential equations.

(i) $\dfrac{d^2y}{dt^2} + 6\dfrac{dy}{dt} + 5y = 0.$

(ii) $\dfrac{d^2y}{dt^2} + 6\dfrac{dy}{dt} + 5y = 15.$

(iii) $\dfrac{d^2y}{dt^2} + 8\dfrac{dy}{dt} + 16y = 32t.$

(i) The differential equation

$$\frac{d^2y}{dt^2} + 6\frac{dy}{dt} + 5y = 0$$

is of type (14.7) and theorem 14.2 may be applied directly. The corresponding quadratic equation is

$$s^2 + 6s + 5 = 0$$

which has solutions $s_1 = -1$ and $s_2 = -5$. Therefore, we require Theorem 14.2 (i) and the general solution is

$$y = A\,\mathrm{e}^{-t} + B\,\mathrm{e}^{-5t}.$$

(ii) To solve the differential equation

$$\frac{\mathrm{d}^2 y}{\mathrm{d}t^2} + 6\frac{\mathrm{d}y}{\mathrm{d}t} + 5y = 15,$$

we require to determine y_{CF} and y_{PI} as defined in Theorem 14.3. $y = y_{\mathrm{CF}}$ is the general solution of the equation obtained when the right-hand side is replaced by zero. Therefore, from (i),

$$y_{\mathrm{CF}} = A\,\mathrm{e}^{-t} + B\,\mathrm{e}^{-5t}.$$

Since the right-hand side of the given differential equation is a constant, we choose a trial solution $y = K$ and substitute $y = K$, $\mathrm{d}y/\mathrm{d}t = 0$ and $\mathrm{d}^2y/\mathrm{d}t^2 = 0$ into the given equation, i.e. $0 + 0 + 5K = 15$. Therefore, $K = 3$ and $y_{\mathrm{PI}} = 3$. Applying Theorem 14.3 the required general solution is

$$y = A\,\mathrm{e}^{-t} + B\,\mathrm{e}^{-5t} + 3.$$

(iii) $$\frac{\mathrm{d}^2 y}{\mathrm{d}t^2} + 8\frac{\mathrm{d}y}{\mathrm{d}t} + 16y = 32t.$$

We apply Theorem 14.2 to the equation obtained when the right-hand side is replaced by zero. The corresponding quadratic equation is

$$s^2 + 8s + 16 = 0$$

which has solutions $s_1 = s_2 = -4$. Therefore, from Theorem 14.2 (ii),

$$y_{\mathrm{CF}} = (A + Bt)\,\mathrm{e}^{-4t}.$$

Since the right-hand side of the given differential equation is a polynomial of degree 1, we choose a trial function $y = K_1 t + K_2$ and substitute

$$y = K_1 t + K_2,$$

$$\frac{\mathrm{d}y}{\mathrm{d}t} = K_1$$

and

$$\frac{\mathrm{d}^2 y}{\mathrm{d}t^2} = 0$$

into the equation, i.e.

$$0 + 8K_1 + 16(K_1 t + K_2) = 32t.$$

To satisfy this equation, we require

$$16K_1 = 32 \quad \text{and} \quad 8K_1 + 16K_2 = 0,$$

i.e.

$$K_1 = 2 \quad \text{and} \quad K_2 = -1$$

and

$$y_{\text{PI}} = 2t - 1$$

Applying Theorem 14.3, the required general solution is

$$y = (A + Bt)\,\mathrm{e}^{-4t} + 2t - 1.$$

Example 14.8

Determine the solution of the differential equation

$$\frac{\mathrm{d}^2 y}{\mathrm{d}x^2} + 2\frac{\mathrm{d}y}{\mathrm{d}x} + 5y = 0,$$

given the initial conditions $y = 1$ when $x = 0$, and $\mathrm{d}y/\mathrm{d}x = 0$ when $x = 0$.

Applying Theorem 14.2 directly, the corresponding quadratic equation is

$$s^2 + 2s + 5 = 0$$

with solutions $s = -1 \pm 2i$ where $i = \sqrt{-1}$.

Therefore, from Theorem 14.2 (iii), the general solution is

$$y = \mathrm{e}^{-x}\,(A \cos 2x + B \sin 2x).$$

The condition $y = 1$ when $x = 0$ requires $A = 1$. To satisfy the second initial condition it is necessary to determine $\mathrm{d}y/\mathrm{d}x$.

$$\frac{\mathrm{d}y}{\mathrm{d}x} = \mathrm{e}^{-x}\,(-2A \sin 2x + 2B \cos 2x - A \cos 2x - B \sin 2x).$$

The condition $\mathrm{d}y/\mathrm{d}x = 0$ when $x = 0$ requires $2B = A$, i.e. $B = 0.5$.

Therefore the required solution is

$$y = \mathrm{e}^{-x}\,(\cos 2x + 0.5 \sin 2x).$$

Example 14.9

Determine the general solution of

$$\frac{\mathrm{d}^2 y}{\mathrm{d}t^2} - y = 3\,\mathrm{e}^{at}$$

for the following cases.

(i) $a = 2$.

(ii) $a = 1$.

(i) $$\frac{d^2y}{dt^2} - y = 3\,e^{2t}$$

The corresponding quadratic equation for the *reduced* differential equation is $s^2 - 1 = 0$ with solutions $s_1 = 1$ and $s_2 = -1$; from Theorem 14.2 (i),

$$y_{CF} = A\,e^t + B\,e^{-t}.$$

We choose a trial function $y = K\,e^{2t}$ and substitute

$$y = K\,e^{2t}.$$

$$\frac{dy}{dt} = 2K\,e^{2t}$$

and

$$\frac{d^2y}{dt^2} = 4K\,e^{2t}$$

into the given differential equation, i.e. $(4K - K)\,e^{2t} = 3\,e^{2t}$ which requires $K = 1$. Therefore, $y_{PI} = e^{2t}$ and the general solution is

$$y = A\,e^t + B\,e^{-t} + e^{2t}.$$

(ii) $$\frac{d^2y}{dt^2} - y = 3\,e^t.$$

$$y_{CF} = A\,e^t + B\,e^{-t}$$

as in (i). It is clear that the trial function $y = K\,e^t$ is *not* the correct form for y_{PI} since e^t is a component of the complementary function y_{CF} (substitution of $y = K\,e^t$ into the given differential equation results in the nonsensical equation $0 = 3\,e^t$). Therefore, we choose a trial function $y = Kt\,e^t$ and substitute

$$y = Kt\,e^t,$$

$$\frac{dy}{dt} = K(t + 1)\,e^t$$

and

$$\frac{d^2y}{dt^2} = K(t + 2)\,e^t$$

into the given differential equation, i.e.

$$K(t + 2)\,e^t - Kt\,e^t = 3\,e^t$$

which reduces to $2K\,e^t = 3\,e^t$. Therefore $K = 1.5$ and $y_{PI} = 1.5t\,e^t$. The required general solution is

$$y = A\,e^t + B\,e^{-t} + 1.5t\,e^t.$$

Example 14.10

An undamped mechanical system of natural frequency w is forced into motion by an oscillating external input of frequency n. The system is modelled by the differential equation

$$\frac{d^2x}{dt^2} + w^2x = w^2 \sin nt$$

with initial conditions $x = dx/dt = 0$ when $t = 0$. Express x in terms of t given that $n \neq w$. What form does the function x_{PI} take when $n = w$.

The corresponding quadratic equation for the reduced differential equation is $s^2 + w^2 = 0$ with solutions $s = \pm iw$. We apply Theorem 14.2 (iii) with $u = 0$, $e^{ut} = 1$, and $x_{CF} = A \cos wt + B \sin wt$.

We choose the trial function $x = K_1 \cos nt + K_2 \sin nt$ and substitution in the differential equation gives

$$-n^2K_1 \cos nt - n^2K_2 \sin nt + w^2K_1 \cos nt + w^2K_2 \sin nt = w^2 \sin nt.$$

Equating the coefficients of cos nt and sin nt we obtain

$$-n^2K_1 + w^2K_1 = 0,$$

$$-n^2K_2 + w^2K_2 = w^2,$$

i.e.

$$K_1 = 0,$$

$$K_2 = \frac{w^2}{w^2 - n^2}$$

and

$$x_{PI} = \frac{w^2}{w^2 - n^2} \sin nt, \quad n \neq w.$$

Therefore the general solution of the given differential equation is

$$x = A \cos wt + B \sin wt + \frac{w^2}{w^2 - n^2} \sin nt.$$

The initial conditions $x = dx/dt = 0$ when $t = 0$ require $A = 0$ and $B = -nw/(w^2 - n^2)$. Therefore,

$$x = \frac{w}{w^2 - n^2} (w \sin nt - n \sin wt).$$

When $n = w$, the trial function must be

$$x = t(K_1 \cos nt + K_2 \sin nt).$$

It is left as an exercise for the reader to show that, when $n = w$, the differential equation has the solution

$$x = 0.5(\sin nt - nt \cos nt).$$

Example 14.11

The motion of a parachutist descending with velocity v m/s (measured positive *upwards*) is assumed to satisfy the linear differential equation

$$100\frac{\mathrm{d}v}{\mathrm{d}t} = -981 - 120v.$$

The altitude of the parachutist measured from sea level is denoted by h m. Write down the differential equation satisfied by h and express h in terms of t, given that $h = 1000$ m when $t = 0$, and $v = -40$ m/s when $t = 0$. Estimate the time taken for the parachutist to reach sea level.

We have $v = \mathrm{d}h/\mathrm{d}t$ and therefore

$$\frac{\mathrm{d}^2h}{\mathrm{d}t^2} + 1.2\frac{\mathrm{d}h}{\mathrm{d}t} = -9.81.$$

To obtain h_{CF} we solve the quadratic equation $s^2 + 1.2s = 0$ to obtain $s_1 = -1.2$ and $s_2 = 0$. Therefore,

$$h_{\mathrm{CF}} = A\,\mathrm{e}^{-1.2t} + B.$$

The trial function $h = K$ is no good (it gives $0 = -9.81$); therefore, we try

$$h = Kt, \quad \frac{\mathrm{d}h}{\mathrm{d}t} = K, \quad \frac{\mathrm{d}^2h}{\mathrm{d}t^2} = 0,$$

i.e.

$$1.2K = -9.81 \quad \text{and} \quad h_{\mathrm{PI}} = -8.175t.$$

Therefore,

$$h = A\,\mathrm{e}^{-1.2t} + B - 8.175t$$

and

$$v = -1.2A\,\mathrm{e}^{-1.2t} - 8.175.$$

To satisfy the initial condition, we require

$$1000 = A + B$$

$$-40 = -1.2A - 8.175.$$

Therefore, $A = 26.52$, $B = 973.48$ (approximately) and

$$h \approx 26.52\,\mathrm{e}^{-1.2t} + 973.48 - 8.175t.$$

The parachutist reaches sea level when $h = 0$, i.e. when $-26.52\,e^{-1.2t} = 973.48 - 8.175t$. This equation may be solved using a suitable numerical technique, but it is clear that the exponential function will be negligible for the value of t which satisfies the equation. Therefore,

$$t \approx \frac{973.48}{8.175} \approx 119 \text{ s}$$

at sea level.

Problems

1. Determine the general solution of the following differential equations.

(i) $\dfrac{d^2y}{dt^2} + 3\dfrac{dy}{dt} + 2y = 0.$

(ii) $4\dfrac{d^2y}{dt^2} + 4\dfrac{dy}{dt} + y = 0.$

(iii) $\dfrac{d^2x}{dt^2} + x = 4.$

(iv) $\dfrac{d^2y}{dx^2} + 2\dfrac{dy}{dx} + 10y = 20\,e^{-2x}.$

2. Solve the following differential equations subject to the given initial conditions.

(i) $\dfrac{d^2y}{dx^2} + y = 3 - 2x^2, \quad y(0) = 7, \quad \dfrac{dy}{dx}(0) = 0.$

(ii) $\dfrac{d^2y}{dt^2} - 4\dfrac{dy}{dt} + 3y = e^{2t}, \quad y(0) = \dfrac{dy}{dt}(0) = 0.$

(iii) $\dfrac{d^2x}{dt^2} + 2\dfrac{dx}{dt} + 2x = \sin t, \quad x(0) = 2, \quad \dfrac{dx}{dt}(0) = -1.$

(iv) $\dfrac{d^2y}{dt^2} - 4\dfrac{dy}{dt} + 3y = e^{t}, \quad y(0) = \dfrac{dy}{dt}(0) = 0.$

3. Determine the constants K and L such that $y = K\,e^x + L\,e^{2x}$ is a solution of the differential equation

$$\frac{d^2y}{dx^2} + 2\frac{dy}{dx} + 5y = e^x + 2\,e^{2x}.$$

Hence write down the general solution of the differential equation.

4. Show that $y = t^2\ e^{-2t}$ is a solution of the differential equation

$$\frac{d^2y}{dt^2} + 4\frac{dy}{dt} + 4y = 2\ e^{-2t}.$$

Hence solve the equation given that $y = 1$ when $t = 0$ and $dy/dt = 0$ when $t = 0$.

5. Apply the theorems of section 14.4 to solve the second-order differential equations given in
(i) Example 14.3 and
(ii) problem 2 of section 14.2.

6. Show that the solution of the differential equation

$$\frac{d^2y}{dt^2} + 2kw\frac{dy}{dt} + w^2y = w^2, \quad k > 0, \quad w > 0.$$

is such that $y \to 1$ as $t \to \infty$. Show further that the solution is oscillatory when $0 < k < 1$.

7. A capacitor C discharges through a circuit of resistance R and inductance L. The charge q on the capacitor satisfies the differential equation

$$L\frac{d^2q}{dt^2} + R\frac{dq}{dt} + \frac{q}{C} = 0.$$

Obtain the condition that the discharge is *just* non-oscillatory (i.e. the circuit is *critically damped*).

8. The *small* oscillations of a pendulum satisfy the differential equation

$$\frac{d^2x}{dt^2} + \frac{g}{l}x = 0, \quad x(0) = x_0, \quad \frac{dx}{dt}(0) = 0,$$

where g and l are constants and x denotes the small angular displacement of the pendulum. Express x in terms of t, and state the period of oscillations.

9. The buying behaviour of the public towards a certain product is modelled approximately by the differential equation

$$\frac{d^2B}{dt^2} + (bp + aq)\frac{dB}{dt} + ab(pq - 1)B = cA,$$

where $B(t)$ is the level of buying, $A(t)$ is the advertising policy, and a, b, c, p and q are all positive constants. Deduce that for constant advertising the buying level tends to a limiting value provided that $pq > 1$. When $p = q = 2, a = b = c = 1$ and $A(t) = 100$, determine the level of buying for $t \geqslant 0$ given that $B = dB/dt = 0$ when $t = 0$.

Answers

1. (i) $y = A\,e^{-t} + B\,e^{-2t}$.
   (ii) $y = (A + Bt)\,e^{-t/2}$.
   (iii) $x = A\cos t + B\sin t + 4$.
   (iv) $y = e^{-x}(A\cos 3x + B\sin 3x) + 2\,e^{-2x}$.

2. (i) $y = 7 - 2x^2$.
   (ii) $y = 0.5\,e^{t} + 0.5\,e^{3t} - e^{2t}$.
   (iii) $x = 1.2\,e^{-t}(2\cos t + \sin t) + 0.2(\sin t - 2\cos t)$.
   (iv) $y = 0.25(e^{3t} - e^{t} - 2t\,e^{t})$.

3. $K = \frac{1}{8}$, $L = \frac{2}{13}$,
   $y = e^{-x}(A\cos 2x + B\sin 2x) + \frac{1}{8}e^{x} + \frac{2}{13}e^{2x}$.

4. $y = (1 + 2t + t^2)\,e^{-2t}$.

7. $CR^2 = 4L$.

8. $x = x_0 \cos\left(\sqrt{\frac{g}{l}}\,t\right)$, $2\pi\sqrt{\frac{l}{g}}$.

9. $B = 100(\frac{1}{3} - \frac{1}{2}e^{-t} + \frac{1}{6}e^{-3t})$.

Further Problems for Chapter 14

1. Use the method of separation of variables to solve the growth (or decay) equation $dy/dt = ky$, $y = y_0$ when $t = 0$.

2. Use the integrating factor technique to solve the growth (or decay) equation.

3. Use a method based on Theorems 14.2 and 14.3 to solve the growth (or decay) equation. (The corresponding algebraic equation is not quadratic but linear.)

4. Write down the general solution of the differential equation $d^3y/dt^3 = 12x$.

5. Solve the differential equation $d^2\theta/dt^2 = 6$ given that $\theta = 0$ when $t = 0$ and $d\theta/dt = 3$ when $t = 0$.

6. Solve the differential equation $dy/dx = \sin x$ given that $y = 0$ when $x = 0$.

7. Determine the solution of the differential equation

$$\frac{dy}{dx} = -\frac{x}{y},$$

which satifies the condition $y = 10$ when $x = 9$.

8. Solve the linear differential equation

$$\frac{dy}{dt} + \frac{y}{t} = 2, \quad y(1) = 2.$$

9. Determine the general solution of the second-order differential equation

$$\frac{d^2 y}{dt^2} + 6\frac{dy}{dt} + 10y = 0.$$

10. Write down the general solution of the differential equation

$$\frac{d^2 y}{dx^2} + 2\frac{dy}{dx} + y = 0.$$

11. Determine the solution of the differential equation

$$\frac{d^2 x}{dt^2} + 10\frac{dx}{dt} + 24x = 0,$$

which satifies the conditions $x = dx/dt = 0$ when $t = 0$.

12. Write down the general solution of

$$\frac{d^2 y}{dt^2} - 4y = 24.$$

13. Obtain the general solution of the equation

$$\frac{d^2 q}{dt^2} + 4q = 8t - 20.$$

14. The motion of a flywheel satisfies the equation

$$I\frac{d^2 \theta}{dt^2} + C\frac{d\theta}{dt} = T,$$

$\theta(0) = d\theta/dt\,(0) = 0$, where θ is the angular displacement of the flywheel and I, C and T are constants. Show that $\theta = Kt$ is a suitable trial function and deduce that, as $t \to \infty$, $d\theta/dt \to T/C$.

Answers

1. $y = y_0\, e^{kt}$.
2. $y = y_0\, e^{kt}$.
3. $y = y_0\, e^{kt}$.

4. $y = \frac{x^4}{2} + Ax^2 + Bx + C.$

5. $\theta = 3t^2 + 3t.$

6. $y = 1 - \cos x.$

7. $x^2 + y^2 = 181.$

8. $y = t + \frac{1}{t}.$

9. $y = e^{-3t}(A \cos t + B \sin t).$

10. $y = (A + Bx)\, e^{-x}$

11. $x = 0.$

12. $y = A\, e^{2t} + B\, e^{-2t} - 6.$

13. $q = A \cos 2t + B \sin 2t + 2t - 5.$

References

Ayres, F. (1972) *Differential Equations*, McGraw-Hill.

Dunning-Davies, J. (1982) *Mathematical Methods for Mathematicians, Physical Scientists and Engineers*, Ellis Horwood.

Johnson, R. M. (1984) *Theory and Applications of Linear Differential and Difference Equations, a System Approach in Engineering*, Ellis Horwood.

Phillips, G. M. & Taylor, P. J. (1973) *Theory and Applications of Numerical Analysis,* Academic Press.

Sweet, M. V. (1984) *Algebra, Geometry and Trigonometry in Science, Engineering and Mathematics*, Ellis Horwood.

Thomas, G. B. & Finney, R. L. (1979) *Calculus and Analytic Geometry*, Addison-Wesley.

Table 1 The derivatives of common functions

$f(x)$	$f'(x)$
x^n	nx^{n-1}
$\sin x$	$\cos x$
$\cos x$	$-\sin x$
$\tan x$	$\sec^2 x$
$\operatorname{cosec} x$	$-\operatorname{cosec} x \cot x$
$\sec x$	$\sec x \tan x$
$\cot x$	$-\operatorname{cosec}^2 x$
e^x	e^x
$\ln x$	$\dfrac{1}{x}$
$\sinh x$	$\cosh x$
$\cosh x$	$\sinh x$
$\tanh x$	$\operatorname{sech}^2 x$
$\operatorname{cosech} x$	$-\operatorname{cosech} x \coth x$
$\operatorname{sech} x$	$-\operatorname{sech} x \tanh x$
$\coth x$	$-\operatorname{cosech}^2 x$
$\sin^{-1} x$	$\dfrac{1}{\sqrt{1-x^2}}$
$\cos^{-1} x$	$-\dfrac{1}{\sqrt{1-x^2}}$
$\tan^{-1} x$	$\dfrac{1}{1+x^2}$
$\sinh^{-1} x$	$\dfrac{1}{\sqrt{x^2+1}}$
$\cosh^{-1} x$	$\dfrac{1}{\sqrt{x^2-1}}$
$\tanh^{-1} x$	$\dfrac{1}{1-x^2}$

Table 2 Standard Integrals

$f(x)$	$\int f(x)\,\mathrm{d}x$
$x^n, n \neq -1$	$\dfrac{x^{n+1}}{n+1}$
$\dfrac{1}{x}$	$\ln \lvert x \rvert$
$\sin x$	$-\cos x$
$\cos x$	$\sin x$
$\sec^2 x$	$\tan x$
$\operatorname{cosec} x \cot x$	$-\operatorname{cosec} x$
$\sec x \tan x$	$\sec x$
$\operatorname{cosec}^2 x$	$-\cot x$
e^x	e^x
$\sinh x$	$\cosh x$
$\cosh x$	$\sinh x$
$\dfrac{1}{\sqrt{a^2 - x^2}}$	$\sin^{-1}\left(\dfrac{x}{a}\right)$
$\dfrac{1}{a^2 + x^2}$	$\dfrac{1}{a}\tan^{-1}\left(\dfrac{x}{a}\right)$
$\dfrac{1}{\sqrt{x^2 + a^2}}$	$\sinh^{-1}\left(\dfrac{x}{a}\right)$
$\dfrac{1}{\sqrt{x^2 - a^2}}$	$\cosh^{-1}\left(\dfrac{x}{a}\right)$
$\dfrac{1}{a^2 - x^2}$	$\dfrac{1}{a}\tanh^{-1}\left(\dfrac{x}{a}\right)$

Note the following logarithmic forms:

$$\sinh^{-1} u = \ln(u + \sqrt{u^2 + 1}),$$

$$\cosh^{-1} u = \ln(u + \sqrt{u^2 - 1}), \qquad u \geqslant 1,$$

$$\tanh^{-1} u = \tfrac{1}{2}\ln\left(\frac{1+u}{1-u}\right), \qquad |u| < 1.$$

Index

Statistics and Operational Research

Editor: B. W. CONOLLY, Professor of Operational Research, Queen Mary College, University of London

Beaumont, G.P.	**Introductory Applied Probability**
Beaumont, G.P.	**Probability and Random Variables**
Conolly, B.W.	**Techniques in Operational Research: Vol. 1, Queueing Systems**
Conolly, B.W.	**Techniques in Operational Research: Vol. 2, Models, Search, Randomization**
Conolly, B.W.	**Lecture Notes in Queueing Systems**
French, S.	**Sequencing and Scheduling: Mathematics of the Job Shop**
French, S.	**Decision Theory: An Introduction to the Mathematics of Rationality**
Griffiths, P. & Hill, I.D.	**Applied Statistics Algorithms**
Hartley, R.	**Linear and Non-linear Programming**
Jolliffe, F.R.	**Survey Design and Analysis**
Jones, A.J.	**Game Theory**
Kemp, K.W.	**Dice, Data and Decisions: Introductory Statistics**
Oliveira-Pinto, F.	**Simulation Concepts in Mathematical Modelling**
Oliveira-Pinto, F. & Conolly, B.W.	**Applicable Mathematics of Non-physical Phenomena**
Ratschek, J. & Rokne, J.	**Computer Methods for the Range of Functions**
Ratschek, J. & Rokne, J.	**New Computer Methods for Global Optimisation**
Schendel, U.	**Introduction to Numerical Methods for Parallel Computers**
Stoodley, K.D.C.	**Applied and Computational Statistics: A First Course**
Stoodley, K.D.C., Lewis, T. & Stainton, C.L.S.	**Applied Statistical Techniques**
Thomas, L.C.	**Games, Theory and Applications**
Whitehead, J.R.	**The Design and Analysis of Sequential Clinical Trials**

Cullen, M.R.	**Linear Models in Biology**
Dunning-Davies, J.	**Mathematical Methods for Mathematicians, Physical Scientists and Engineers**
Eason, G., Coles, C.W. & Gettinby, G.	**Mathematics and Statistics for the Bio-sciences**
El-Jai, A. & Pritchard, A.J.	**Control Sensors in Distributed Systems**
Exton, H.	**Handbook of Hypergeometric Integrals**
Exton, H.	**Multiple Hypergeometric Functions and Applications**
Exton, H.	***q*-Hypergeometric Functions and Applications**
Faux, I.D. & Pratt, M.J.	**Computational Geometry for Design and Manufacture**
Firby, P.A. & Gardiner, C.F.	**Surface Topology**
Gardiner, C.F.	**Modern Algebra**
Gardiner, C.F.	**Algebraic Structures: with Applications**
Gasson, P.C.	**Geometry of Spatial Forms**
Goodbody, A.M.	**Cartesian Tensors**
Goult, R.J.	**Applied Linear Algebra**
Graham, A.	**Kronecker Products and Matrix Calculus: with Applications**
Graham, A.	**Matrix Theory and Applications for Engineers and Mathematicians**
Graham, A.	**Nonnegative Matrices and Applicable Topics in Linear Algebra**
Griffel, D.H.	**Applied Functional Analysis**
Griffel, D.H.	**Linear Algebra**
Hanyga, A.	**Mathematical Theory of Non-linear Elasticity**
Harris, D.J.	**Mathematics for Business, Management and Economics**
Hoskins, R.F.	**Generalised Functions**
Hoskins, R.F.	**Standard and Non-standard Analysis**
Hunter, S.C.	**Mechanics of Continuous Media, 2nd (Revised) Edition**
Huntley, I. & Johnson, R.M.	**Linear and Nonlinear Differential Equations**
Jaswon, M.A. & Rose, M.A.	**Crystal Symmetry: The Theory of Colour Crystallography**
Johnson, R.M.	**Theory and Applications of Linear Differential and Difference Equations**
Johnson, R.M.	**Calculus: Theory and Applications in Technology and the Physical and Life Sciences**
Jones, R.H. & Steele, N.C.	**Mathematics of Communication**
Kelly, J.C.	**Abstract Algebra**
Kim, K.H. & Roush, F.W.	**Applied Abstract Algebra**
Kim, K.H. & Roush, F.W.	**Team Theory**
Kosinski, W.	**Field Singularities and Wave Analysis in Continuum Mechanics**
Krishnamurthy, V.	**Combinatorics: Theory and Applications**
Lindfield, G. & Penny, J.E.T.	**Microcomputers in Numerical Analysis**
Lord, E.A. & Wilson, C.B.	**The Mathematical Description of Shape and Form**
Malik, M., Riznichenko, G.Y. & Rubin, A.B.	**Biological Electron Transport Processes and their Computer Simulation**
Marichev, O.I.	**Integral Transforms of Higher Transcendental Functions**
Massey, B.S.	**Measures in Science and Engineering**
Meek, B.L. & Fairthorne, S.	**Using Computers**
Mikolas, M.	**Real Functions and Orthogonal Series**
Moore, R.	**Computational Functional Analysis**
Müller-Pfeiffer, E.	**Spectral Theory of Ordinary Differential Operators**
Murphy, J.A. & McShane, B.	**Computation in Numerical Analysis**
Nonweiler, T.R.F.	**Computational Mathematics: An Introduction to Numerical Approximation**
Ogden, R.W.	**Non-linear Elastic Deformations**
Oldknow, A.	**Microcomputers in Geometry**
Oldknow, A. & Smith, D.	**Learning Mathematics with Micros**
O'Neill, M.E. & Chorlton, F.	**Ideal and Incompressible Fluid Dynamics**
O'Neill, M.E. & Chorlton, F.	**Viscous and Compressible Fluid Dynamics**
Page, S. G.	**Mathematics: A Second Start**
Prior, D. & Moscardini, A.O.	**Model Formulation Analysis**
Rankin, R.A.	**Modular Forms**
Scorer, R.S.	**Environmental Aerodynamics**
Smith, D.K.	**Network Optimisation Practice: A Computational Guide**
Srivastava, H.M. & Karlsson, P.W.	**Multiple Gaussian Hypergeometric Series**
Srivastava, H.M. & Manocha, H.L.	**A Treatise on Generating Functions**
Shivamoggi, B.K.	**Stability of Parallel Gas Flows**
Stirling, D.S.G.	**Mathematical Analysis**
Sweet, M.V.	**Algebra, Geometry and Trigonometry in Science, Engineering and Mathematics**
Temperley, H.N.V. & Trevena, D.H.	**Liquids and Their Properties**
Temperley, H.N.V.	**Graph Theory and Applications**
Thom, R.	**Mathematical Models of Morphogenesis**
Toth, G.	**Harmonic and Minimal Maps**
Townend, M. S.	**Mathematics in Sport**
Towend, M.S. & Pountney, D.C.	**Computer-Aided Engineering Mathematics**
Twizell, E.H.	**Computational Methods for Partial Differential Equations**
Twizel, E.H.	**numerical Methods in the Biomedical Sciences**
Virice, A. and Morris, C.	**Mathematics for Computer Studies**
Walton, K., Marshall, J., Gorecki, H. & Konytowski, A.	**Control Theory for Time Delay Systems**
Wheeler, R.F.	**Rethinking Mathematical Concepts**
Willmore, T.J.	**Total Curvature in Riemannian Geometry**
Willmore, T.J. & Hitchin, N.	**Global Riemannian Geometry**
Wojtynski, W.	**Lie Groups and Lie Algebras**